Looking for Pythagoras

The Pythagorean Theorem

Teacher's Edition

Glenda Lappan
James T. Fey
William M. Fitzgerald
Susan N. Friel
Elizabeth Difanis Phillips

Developed at Michigan State University

DALE SEYMOUR PUBLICATIONS®

Connected Mathematics™ was developed at Michigan State University with financial support from the Michigan State University Office of the Provost, Computing and Technology, and the College of Natural Science.

This material is based upon work supported by the National Science Foundation under Grant No. MDR 9150217.

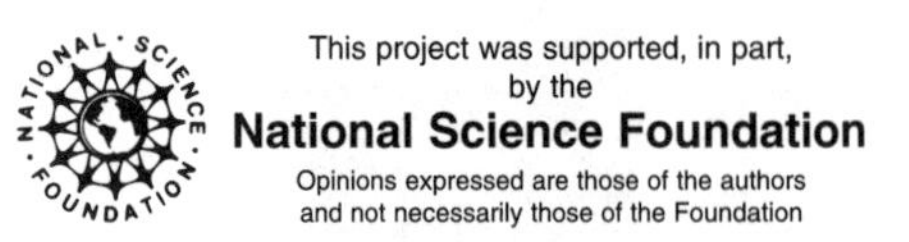

The Michigan State University authors and administration have agreed that all MSU royalties arising from this publication will be devoted to purposes supported by the Department of Mathematics and the MSU Mathematics Education Enrichment Fund.

This book is published by Dale Seymour Publications®, an imprint of Addison Wesley Longman, Inc.

Managing Editor: Catherine Anderson
Project Editor: Stacey Miceli
Book Editor: Mali Apple
ESL Consultant: Nancy Sokol Green
Production/Manufacturing Director: Janet Yearian
Production/Manufacturing Coordinator: Claire Flaherty
Design Manager: John F. Kelly
Photo Editor: Roberta Spieckerman
Design: Don Taka
Composition: London Road Design, Palo Alto, CA
Illustrations: Pauline Phung, Margaret Copeland, Ray Godfrey
Cover: Ray Godfrey

Photo Acknowledgements: 5 © Margot Granitsas/The Image Works; 10 © Robert Llewellyn/Superstock, Inc.; 13 © Bob Daemmrich/The Image Works; 14 © Laima Druskis/Stock, Boston; 30 © ET Archive/Superstock, Inc.; 41 © Lambert/Archive Photos

Order number 21476
ISBN 1-57232-181-4

1 2 3 4 5 6 7 8 9 10-ML-01 00 99 98 97

The Connected Mathematics Project Staff

Project Directors

James T. Fey
University of Maryland

William M. Fitzgerald
Michigan State University

Susan N. Friel
University of North Carolina at Chapel Hill

Glenda Lappan
Michigan State University

Elizabeth Difanis Phillips
Michigan State University

Project Manager

Kathy Burgis
Michigan State University

Technical Coordinator

Judith Martus Miller
Michigan State University

Collaborating Teachers/Writers

Mary K. Bouck
Portland, Michigan

Jacqueline Stewart
Okemos, Michigan

Curriculum Development Consultants

David Ben-Chaim
Weizmann Institute

Alex Friedlander
Weizmann Institute

Eleanor Geiger
University of Maryland

Jane Mitchell
University of North Carolina at Chapel Hill

Anthony D. Rickard
Alma College

Evaluation Team

Mark Hoover
Michigan State University

Diane V. Lambdin
Indiana University

Sandra K. Wilcox
Michigan State University

Judith S. Zawojewski
National-Louis University

Graduate Assistants

Scott J. Baldridge
Michigan State University

Angie S. Eshelman
Michigan State University

M. Faaiz Gierdien
Michigan State University

Jane M. Keiser
Indiana University

Angela S. Krebs
Michigan State University

James M. Larson
Michigan State University

Ronald Preston
Indiana University

Tat Ming Sze
Michigan State University

Sarah Theule-Lubienski
Michigan State University

Jeffrey J. Wanko
Michigan State University

Field Test Production Team

Katherine Oesterle
Michigan State University

Stacey L. Otto
University of North Carolina at Chapel Hill

Teacher/Assessment Team

Kathy Booth
Waverly, Michigan

Anita Clark
Marshall, Michigan

Julie Faulkner
Traverse City, Michigan

Theodore Gardella
Bloomfield Hills, Michigan

Yvonne Grant
Portland, Michigan

Linda R. Lobue
Vista, California

Suzanne McGrath
Chula Vista, California

Nancy McIntyre
Troy, Michigan

Mary Beth Schmitt
Traverse City, Michigan

Linda Walker
Tallahassee, Florida

Software Developer

Richard Burgis
East Lansing, Michigan

Development Center Directors

Nicholas Branca
San Diego State University

Dianne Briars
Pittsburgh Public Schools

Frances R. Curcio
New York University

Perry Lanier
Michigan State University

J. Michael Shaughnessy
Portland State University

Charles Vonder Embse
Central Michigan University

Field Test Coordinators

Michelle Bohan
Queens, New York

Melanie Branca
San Diego, California

Alecia Devantier
Shepherd, Michigan

Jenny Jorgensen
Flint, Michigan

Sandra Kralovec
Portland, Oregon

Sonia Marsalis
Flint, Michigan

William Schaeffer
Pittsburgh, Pennsylvania

Karma Vince
Toledo, Ohio

Virginia Wolf
Pittsburgh, Pennsylvania

Shirel Yaloz
Queens, New York

Student Assistants

Laura Hammond
David Roche
Courtney Stoner
Jovan Trpovski
Julie Valicenti
Michigan State University

Advisory Board

Joseph Adney
Michigan State University (Emeritus)

Charles Allan
Michigan Department of Education

Mary K. Bouck
Portland Public Schools
Portland, Michigan

C. Stuart Brewster
Palo Alto, California

Anita Clark
Marshall Public Schools
Marshall, Michigan

David Doherty
GMI Engineering and Management Institute
Flint, Michigan

Kay Gilliland
EQUALS
Berkeley, California

David Green
GMI Engineering and Management Institute
Flint, Michigan

Henry Heikkinen
University of Northern Colorado
Greeley, Colorado

Anita Johnston
Jackson Community College
Jackson, Michigan

Elizabeth M. Jones
Lansing School District
Lansing, Michigan

Jim Landwehr
AT&T Bell Laboratories
Murray Hill, New Jersey

Peter Lappan
Michigan State University

Steven Leinwand
Connecticut Department of Education

Nancy McIntyre
Troy Public Schools
Troy, Michigan

Valerie Mills
Ypsilanti Public Schools
Ypsilanti, Michigan

David S. Moore
Purdue University
West Lafayette, Indiana

Ralph Oliva
Texas Instruments
Dallas, Texas

Richard Phillips
Michigan State University

Jacob Plotkin
Michigan State University

Dawn Pysarchik
Michigan State University

Rheta N. Rubenstein
University of Windsor
Windsor, Ontario, Canada

Susan Jo Russell
TERC
Cambridge, Massachusetts

Marjorie Senechal
Smith College
Northampton, Massachusetts

Sharon Senk
Michigan State University

Jacqueline Stewart
Okemos School District
Okemos, Michigan

Uri Treisman
University of Texas
Austin, Texas

Irvin E. Vance
Michigan State University

Linda Walker
Tallahassee Public Schools
Tallahassee, Florida

Gail Weeks
Northville Public Schools
Northville, Michigan

Pilot Teachers

California

National City

Laura Chavez
National City Middle School

Ruth Ann Duncan
National City Middle School

Sonia Nolla
National City Middle School

San Diego

James Ciolli
Los Altos Elementary School

Chula Vista

Agatha Graney
Hilltop Middle School

Suzanne McGrath
Eastlake Elementary School

Toni Miller
Hilltop Middle School

Lakeside

Eva Hollister
Tierra del Sol Middle School

Vista

Linda LoBue
Washington Middle School

Illinois

Evanston

Marlene Robinson
Baker Demonstration School

Michigan

Bloomfield Hills

Roxanne Cleveland
Bloomfield Hills Middle School

Constance Kelly
Bloomfield Hills Middle School

Tim Loula
Bloomfield Hills Middle School

Audrey Marsalese
Bloomfield Hills Middle School

Kara Reid
Bloomfield Hills Middle School

Joann Schultz
Bloomfield Hills Middle School

Flint

Joshua Coty
Holmes Middle School

Brenda Duckett-Jones
Brownell Elementary School

Lisa Earl
Holmes Middle School

Anne Heidel
Holmes Middle School

Chad Meyers
Brownell Elementary School

Greg Mickelson
Holmes Middle School

Rebecca Ray
Holmes Middle School

Patricia Wagner
Holmes Middle School

Greg Williams
Gundry Elementary School

Lansing

Susan Bissonette
Waverly Middle School

Kathy Booth
Waverly East Intermediate School

Carole Campbell
Waverly East Intermediate School

Gary Gillespie
Waverly East Intermediate School

Denise Kehren
Waverly Middle School

Virginia Larson
Waverly East Intermediate School

Kelly Martin
Waverly Middle School

Laurie Metevier
Waverly East Intermediate School

Craig Paksi
Waverly East Intermediate School

Tony Pecoraro
Waverly Middle School

Helene Rewa
Waverly East Intermediate School

Arnold Stiefel
Waverly Middle School

Portland

Bill Carlton
Portland Middle School

Kathy Dole
Portland Middle School

Debby Flate
Portland Middle School

Yvonne Grant
Portland Middle School

Terry Keusch
Portland Middle School

John Manzini
Portland Middle School

Mary Parker
Portland Middle School

Scott Sandborn
Portland Middle School

Shepherd

Steve Brant
Shepherd Middle School

Marty Brock
Shepherd Middle School

Cathy Church
Shepherd Middle School

Ginny Crandall
Shepherd Middle School

Craig Ericksen
Shepherd Middle School

Natalie Hackney
Shepherd Middle School

Bill Hamilton
Shepherd Middle School

Julie Salisbury
Shepherd Middle School

Sturgis

Sandra Allen
Eastwood Elementary School

Margaret Baker
Eastwood Elementary School

Steven Baker
Eastwood Elementary School

Keith Barnes
Sturgis Middle School

Wilodean Beckwith
Eastwood Elementary School

Darcy Bird
Eastwood Elementary School

Bill Dickey
Sturgis Middle School

Ellen Eisele
Sturgis Middle School

James Hoelscher
Sturgis Middle School

Richard Nolan
Sturgis Middle School

J. Hunter Raiford
Sturgis Middle School

Cindy Sprowl
Eastwood Elementary School

Leslie Stewart
Eastwood Elementary School

Connie Sutton
Eastwood Elementary School

Traverse City

Maureen Bauer
Interlochen Elementary School

Ivanka Berskshire
East Junior High School

Sarah Boehm
Courtade Elementary School

Marilyn Conklin
Interlochen Elementary School

Nancy Crandall
Blair Elementary School

Fran Cullen
Courtade Elementary School

Eric Dreier
Old Mission Elementary School

Lisa Dzierwa
Cherry Knoll Elementary School

Ray Fouch
West Junior High School

Ed Hargis
Willow Hill Elementary School

Richard Henry
West Junior High School

Dessie Hughes
Cherry Knoll Elementary School

Ruthanne Kladder
Oak Park Elementary School

Bonnie Knapp
West Junior High School

Sue Laisure
Sabin Elementary School

Stan Malaski
Oak Park Elementary School

Jody Meyers
Sabin Elementary School

Marsha Myles
East Junior High School

Mary Beth O'Neil
Traverse Heights Elementary School

Jan Palkowski
East Junior High School

Karen Richardson
Old Mission Elementary School

Kristin Sak
Bertha Vos Elementary School

Mary Beth Schmitt
East Junior High School

Mike Schrotenboer
Norris Elementary School

Gail Smith
Willow Hill Elementary School

Karrie Tufts
Eastern Elementary School

Mike Wilson
East Junior High School

Tom Wilson
West Junior High School

Minnesota

Minneapolis

Betsy Ford
Northeast Middle School

New York

East Elmhurst

Allison Clark
Louis Armstrong Middle School

Dorothy Hershey
Louis Armstrong Middle School

J. Lewis McNeece
Louis Armstrong Middle School

Rossana Perez
Louis Armstrong Middle School

Merna Porter
Louis Armstrong Middle School

Marie Turini
Louis Armstrong Middle School

North Carolina

Durham

Everly Broadway
Durham Public Schools

Thomas Carson
Duke School for Children

Mary Hebrank
Duke School for Children

Bill O'Connor
Duke School for Children

Ruth Pershing
Duke School for Children

Peter Reichert
Duke School for Children

Elizabeth City

Rita Banks
Elizabeth City Middle School

Beth Chaundry
Elizabeth City Middle School

Amy Cuthbertson
Elizabeth City Middle School

Deni Dennison
Elizabeth City Middle School

Jean Gray
Elizabeth City Middle School

John McMenamin
Elizabeth City Middle School

Nicollette Nixon
Elizabeth City Middle School

Malinda Norfleet
Elizabeth City Middle School

Joyce O'Neal
Elizabeth City Middle School

Clevie Sawyer
Elizabeth City Middle School

Juanita Shannon
Elizabeth City Middle School

Terry Thorne
Elizabeth City Middle School

Rebecca Wardour
Elizabeth City Middle School

Leora Winslow
Elizabeth City Middle School

Franklinton

Susan Haywood
Franklinton Elementary School

Clyde Melton
Franklinton Elementary School

Louisburg

Lisa Anderson
Terrell Lane Middle School

Jackie Frazier
Terrell Lane Middle School

Pam Harris
Terrell Lane Middle School

Ohio

Toledo

Bonnie Bias
Hawkins Elementary School

Marsha Jackish
Hawkins Elementary School

Lee Jagodzinski
DeVeaux Junior High School

Norma J. King
Old Orchard Elementary School

Margaret McCready
Old Orchard Elementary School

Carmella Morton
DeVeaux Junior High School

Karen C. Rohrs
Hawkins Elementary School

Marie Sahloff
DeVeaux Junior High School

L. Michael Vince
McTigue Junior High School

Brenda D. Watkins
Old Orchard Elementary School

Oregon

Canby

Sandra Kralovec
Ackerman Middle School

Portland

Roberta Cohen
Catlin Gabel School

David Ellenberg
Catlin Gabel School

Sara Normington
Catlin Gabel School

Karen Scholte-Arce
Catlin Gabel School

West Linn

Marge Burack
Wood Middle School

Tracy Wygant
Athey Creek Middle School

Pennsylvania

Pittsburgh

Sheryl Adams
Reizenstein Middle School

Sue Barie
Frick International Studies Academy

Suzie Berry
Frick International Studies Academy

Richard Delgrosso
Frick International Studies Academy

Janet Falkowski
Frick International Studies Academy

Joanne George
Reizenstein Middle School

Harriet Hopper
Reizenstein Middle School

Chuck Jessen
Reizenstein Middle School

Ken Labuskes
Reizenstein Middle School

Barbara Lewis
Reizenstein Middle School

Sharon Mihalich
Reizenstein Middle School

Marianne O'Connor
Frick International Studies Academy

Mark Sammartino
Reizenstein Middle School

Washington

Seattle

Chris Johnson
University Preparatory Academy

Rick Purn
University Preparatory Academy

Contents

Overview

In *Looking for Pythagoras,* students explore two important new ideas: the Pythagorean Theorem and irrational numbers. In the process of solving the problems in this unit, students also review and make connections among the concepts of area, distance, slope, and rational numbers.

Students begin the unit by finding the distance between dots on a grid. They explore the areas of figures drawn on a dot grid, and they find relationships among the length of a side of a square, square roots, and irrational numbers. They find that the side lengths of some squares are irrational numbers. Then, students discover the Pythagorean relationship through an exploration of squares drawn on the sides of a right triangle.

The coordinate system makes all these investigations possible. A coordinate system applied to a rectangular array of dots facilitates locating positions, calculating distances, and finding slopes of lines. We can extend the dot grid to include points between the dots that are visible and create a coordinate system on which we can locate any position in the plane. In this unit, we use only integer scales on the axes, which means that each dot has integer coordinates. The breakthrough for students comes when they can find distances on the integer coordinate system, such as $\sqrt{2}$, that are irrational.

Later in the unit, students discover that a line with an irrational slope that passes through the origin will not pass through any dot on a dot grid. This is because every dot on a grid is associated with a rational slope; irrational slopes, which cannot be expressed as a ratio of integers, are not associated with dots on the grid.

Students' work in this unit develops a fundamentally important relationship connecting geometry and algebra: the Pythagorean Theorem. The presentation of ideas in the unit reflects the historical development of the concept of irrational numbers. The need for such numbers was recognized by early Greek mathematicians as they searched for a ratio of integers to represent the length of the sides of a square with certain given areas, such as 2 square units. The square root of 2 is an irrational number, meaning that it cannot be written as a ratio of two integers.

Finding Slope, Area, and Distance

In this unit, students explore lines and figures drawn on dot grids. Line segments *AB* and *CD* below have endpoints *A* ($^-4$, 2), *B* ($^-1$, 5), *C* (2, $^-2$), and *D* (5, 4). On a dot grid, it is fairly easy to calculate the slope of a line as the ratio of the vertical change to the horizontal change between two points on the line. The slope of line *AB* is 1; the slope of line *CD* is 2.

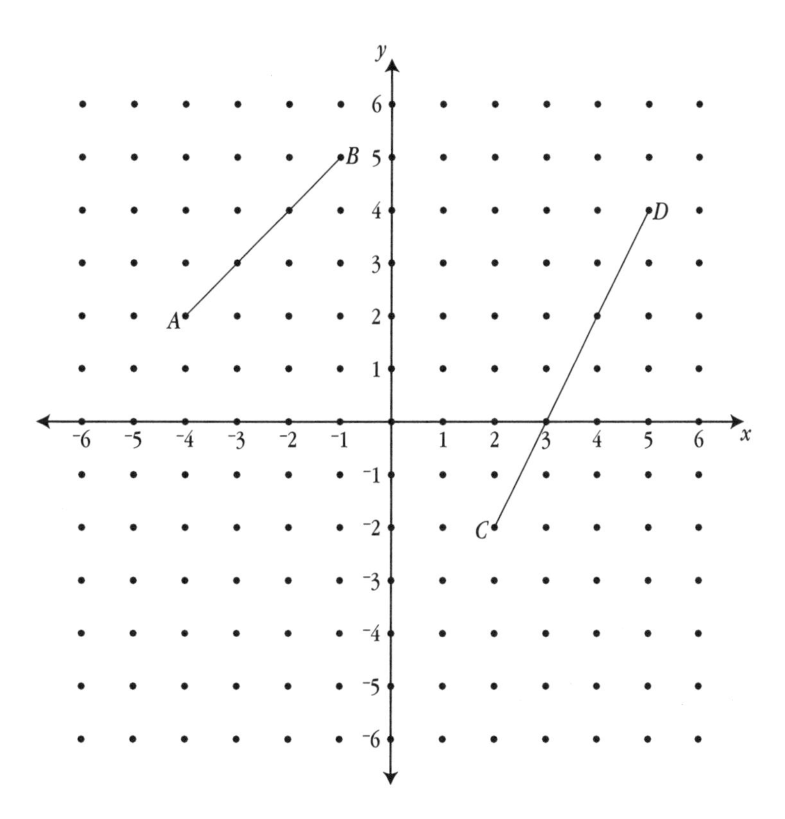

Students also find areas of plane figures drawn on dot grids. One common method used to calculate the area of a figure is to subdivide the figure and add the areas of the component shapes. A second common method is to enclose the figure in a rectangle and subtract the areas of the shapes that lie outside the figure from the area of the rectangle. The area of the figure on the following page is found with each method.

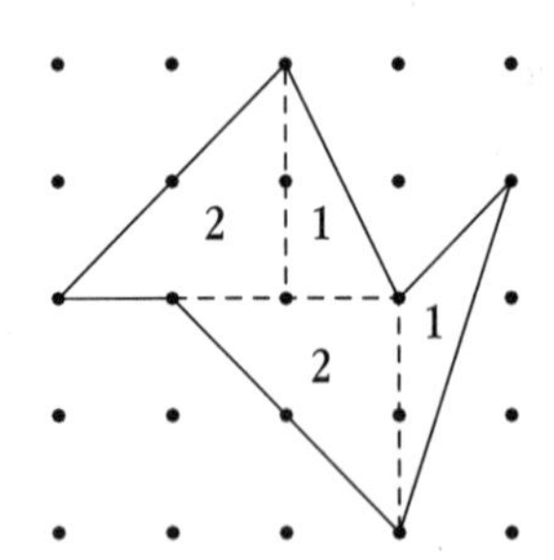

Subdivide to find the area:
$2 + 2 + 1 + 1 = 6$

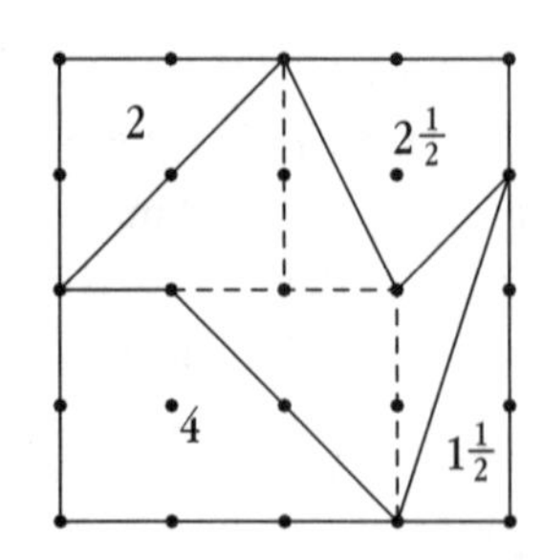

Enclose in a square to find the area:
$16 - (4 + 2 + 2\frac{1}{2} + 1\frac{1}{2}) = 6$

In Investigation 2, students are asked to find all the different areas of squares that can be drawn on a 5-dot-by-5-dot grid. There are eight possible squares, four "upright" and four "tilted."

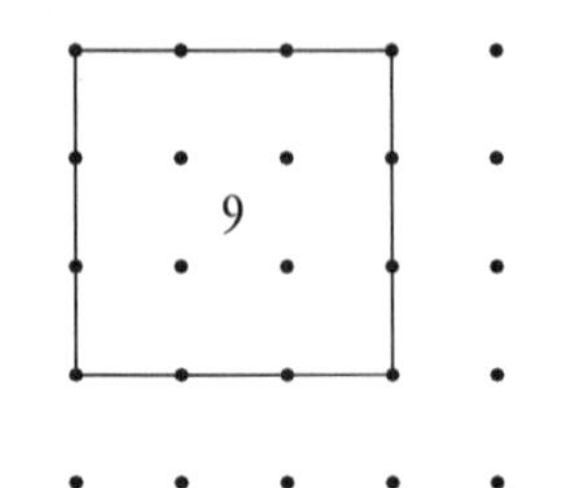

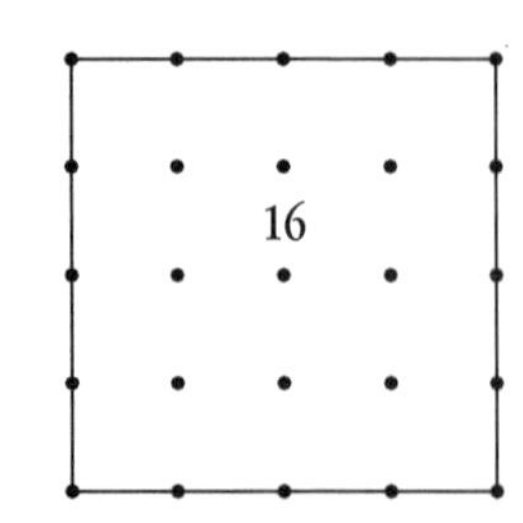

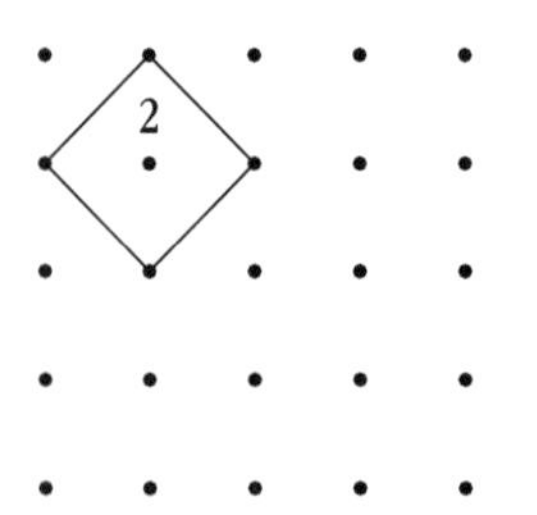

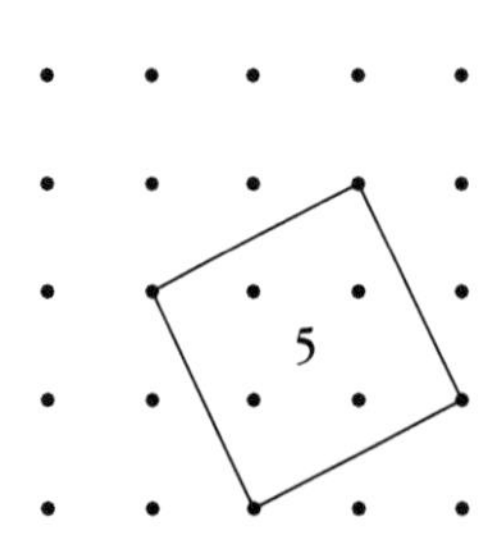

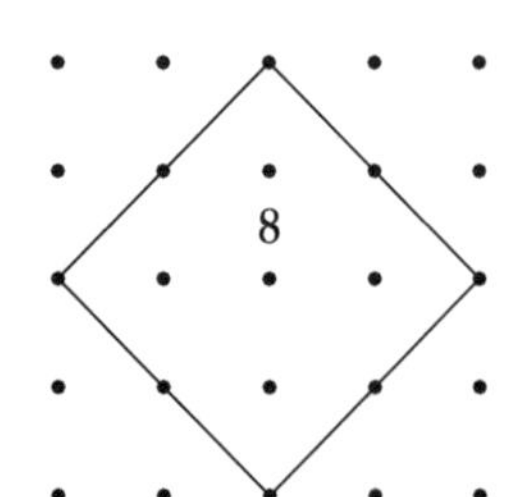

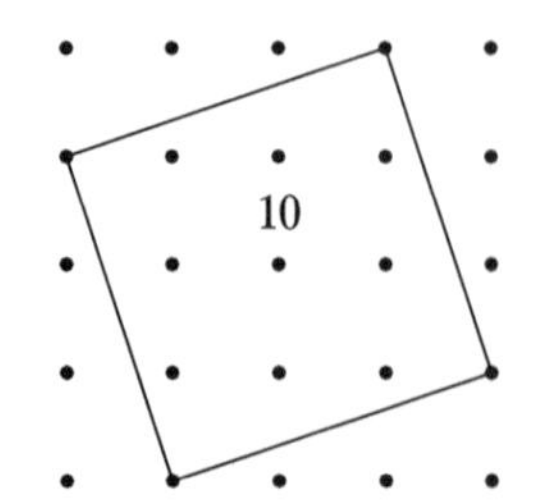

The lengths of the sides of the squares above are 1, 2, 3, 4, $\sqrt{2}$, $\sqrt{5}$, $\sqrt{8}$, and $\sqrt{10}$. If the area of a square is known, its side length is easy to determine: it is the number whose square is the area. The fact that some of these lengths are not whole numbers prompts the introduction of the $\sqrt{\ }$ symbol. By measuring these lengths with a ruler and comparing the measurements to the results of computing the square roots with their calculators, students begin to get a feeling for these numbers that they cannot be expressed as either a terminating or a repeating decimal.

Finding the areas of squares leads students to a method for finding the distance between two points. The distance between two dots on a dot grid is the length of the line segment drawn between them. To find this length, students can draw a square with the segment as one side. The distance between the two dots is the square root of the area of the square.

To find all the different line segment lengths that can be drawn on a 5-dot-by-5-dot grid, the grid would have to be extended to enable the drawing of the squares that are associated with those lengths. For example, the line segment shown below is the side of a square with an area of 25 square units, so the segment has length $\sqrt{25}$, or 5.

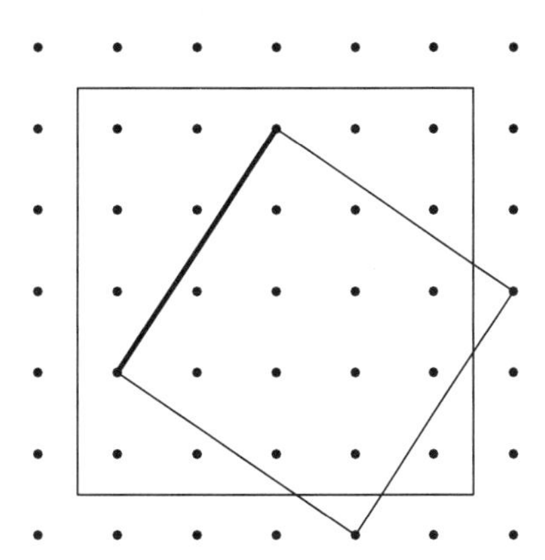

Discovering and Using the Pythagorean Theorem

Once students are comfortable with finding the length of a line segment by thinking of the segment as the side of a square, they investigate the patterns among the areas of the three squares that can be drawn on the sides of a right triangle. For example:

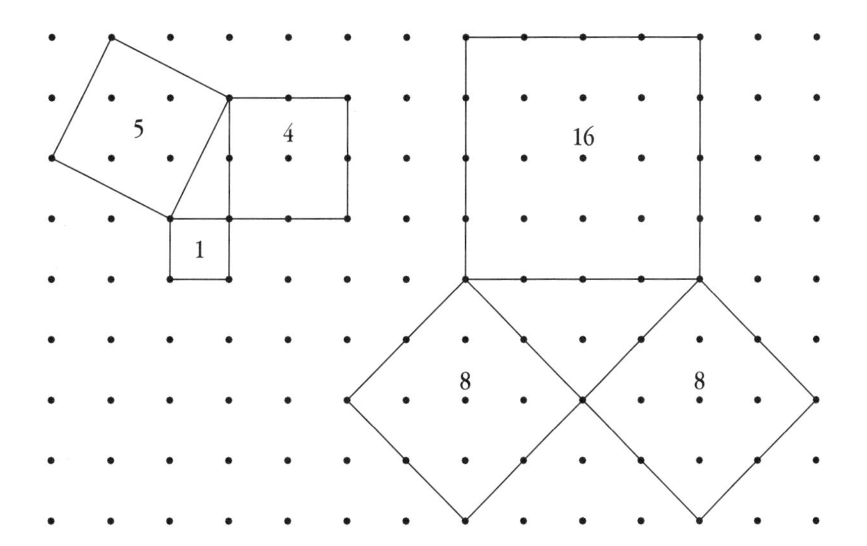

This observation leads students to discover the Pythagorean Theorem.

The Pythagorean Theorem is a remarkable statement with many applications that connect the concepts of line segment lengths, squares, and right angles. The theorem states: *The sum of the areas of the squares on the legs of a right triangle is equal to the area of the square on the hypotenuse.* Symbolically, $a^2 + b^2 = c^2$.

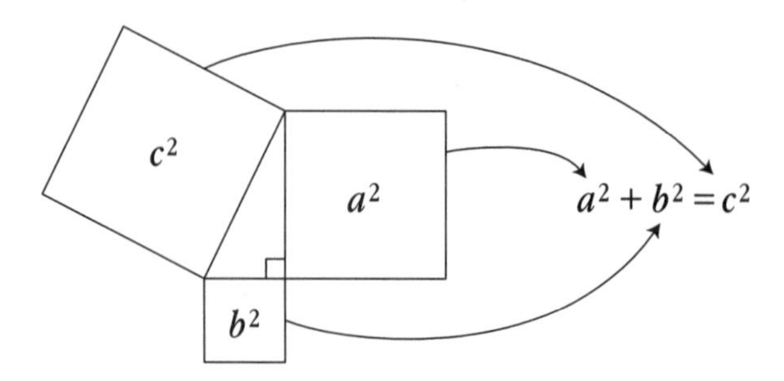

Students use the Pythagorean Theorem to find the distance between two dots on a dot grid. The length of a horizontal or vertical line segment drawn on a dot grid can be found by counting the distance directly. However, if the segment is not vertical or horizontal, it is always possible to treat it as the hypotenuse of a right triangle that has legs that *are* vertical and horizontal and to find the length of the hypotenuse—and thus the distance between the dots—using the Pythagorean Theorem.

For example, to find the length of line segment *AB* below, draw a right triangle with segment *AB* as the hypotenuse. Calculate the areas of the squares on the legs of the triangle (4 square units each), sum the squares (8 square units), and take the square root of this sum. The length of *AB* is thus $\sqrt{8}$.

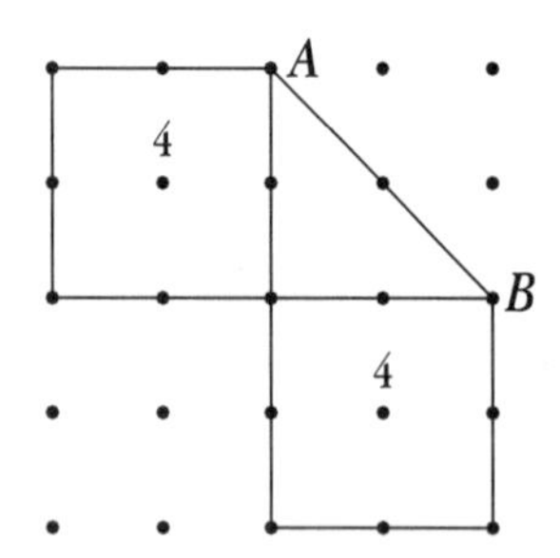

The converse of the Pythagorean Theorem can be used to prove that a triangle is a right triangle. The converse states: *If a, b, and c are the lengths of the sides of a triangle and $a^2 + b^2 = c^2$, then the triangle is a right triangle.*

An interesting by-product of the Pythagorean Theorem is the concept of *Pythagorean triples,* sets of numbers that satisfy the relationship $a^2 + b^2 = c^2$. In this unit, students discover that finding Pythagorean triples means finding two square numbers whose sum is also a square number. Multiples of one triple will generate countless others, such as (3, 4, 5), (6, 8, 10), and (9, 12, 15); or (5, 12, 13), (10, 24, 26), and (15, 36, 39).

Number Systems

New number systems are created when a problem arises that cannot be answered within the system currently in use, or when inconsistencies arise that can be taken care of only by expanding the domain of numbers in the system. The historical "discovery" of a new number system in response to a need is reflected in the number sets students use in grades K–12. Elementary students begin with the *counting numbers.* Then, zero is added to the system to create the set of whole numbers. Later, students discover that negative numbers are needed to give meaning in certain contexts, such as temperature. Now they have the number system called the *integers.* In elementary and middle school, students learn about fractions and situations that make fractions helpful. Their number world has been expanded to the set of *rational numbers.* In this unit, students encounter contexts in which the need for *irrational numbers* arises. The set of rational numbers and the set of irrational numbers compose the set of real numbers. The diagram below is one way to represent these sets of numbers.

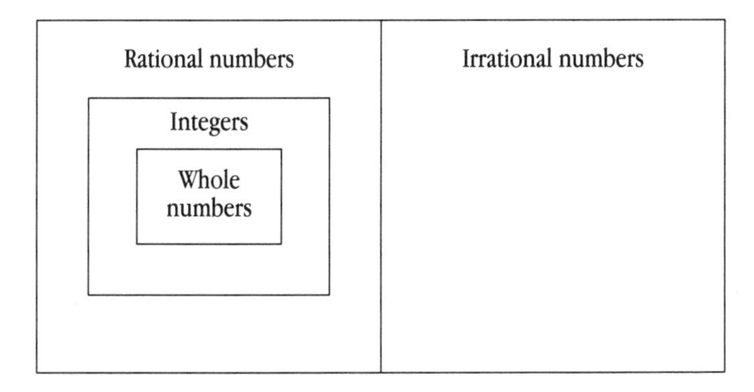

Rational numbers: Numbers that can be written as a ratio of integers, such as $\frac{2}{3}$, $\frac{-7}{5}$, $\frac{4}{1}$, 0.6, and 0.3333 . . . ; in other words, integers, terminating decimals, and repeating decimals.

Integers: {. . . , –3, –2, –1, 0, 1, 2, 3, . . .}

Whole numbers: {0, 1, 2, 3, . . .}

Irrational numbers: Numbers that cannot be written as a ratio of integers, such as π, $\sqrt{7}$, and 0.12131415161718 . . . ; in other words, infinite, nonrepeating decimals.

Exploring Irrational Numbers

In *Looking for Pythagoras,* students are gradually introduced to irrational numbers. They discover the need for irrational numbers by trying to measure the lengths of oblique, or tilted, lines drawn on dot grids. They find such numbers as $\sqrt{2}$, $\sqrt{3}$, and $\sqrt{5}$, which cannot be expressed as ratios of integers. They examine patterns in the decimal representations of fractions, or rational numbers, and find that the decimals either terminate or repeat. For example, $\frac{1}{5} = 0.2$ (a terminating decimal) and $\frac{1}{3} = 0.333\ldots$ (a repeating decimal).

Numbers such as $\sqrt{2}$, $\sqrt{3}$, $\sqrt{5}$, and π cannot be expressed as repeating decimals or terminating decimals. Students create line segments with these lengths and then locate the lengths on a number line. For example, $\sqrt{2}$ is the length of the hypotenuse of a right triangle whose legs have length 1. This procedure helps students to estimate the size of these irrational numbers.

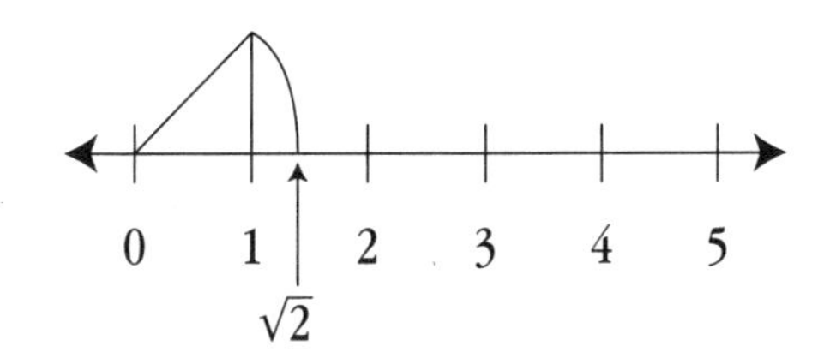

Writing Repeating Decimals as Fractions

One method for converting a repeating decimal to a fraction involves solving an equation. To convert 12.312312 . . . to a fraction, for example, call the unknown fraction N. Thus, $N = 12.312312 \ldots$. Multiply both sides of the equation by 1000 (the power of 10 that moves a complete repeating group to the left of the decimal point), which gives $1000N = 12312.312312 \ldots$. Then, subtract the first equation from the second, which gives $999N = 12{,}300$. Therefore, $N = \frac{12{,}300}{999}$, or $12\frac{312}{999}$.

In this unit, students will recognize that the decimal equivalents of fractions with denominators of 9, 99, 999, and so on, display interesting patterns that can be used to write repeating decimals as fractions. For example, all decimals with a repeating part of one digit, such as 0.111 . . . and 0.222 . . . , can be written as a fraction with 9 in the denominator and the repeated digit in the numerator, such as $\frac{1}{9}$ and $\frac{2}{9}$. Decimals with a repeating part of two digits, such as 0.010101 . . . and 0.121212 . . . , can be written as a fraction with 99 in the denominator and the repeated digits in the numerator, such as $\frac{1}{99}$ and $\frac{12}{99}$.

Proof That $\sqrt{2}$ Is Irrational

In high school, students may prove that $\sqrt{2}$ is not a rational number. Its irrationality can be proved in an interesting way—a proof by contradiction. The proof is given here for the teacher's information.

Assume $\sqrt{2}$ is rational. Then, there exist positive integers p and q such that $\sqrt{2} = \frac{p}{q}$. So, $\sqrt{2}q = p$. Squaring both sides gives $2q^2 = p^2$. The prime factorizations of q^2 and p^2 must each have an even number of factors. (If q has an odd number of factors, then q^2 has an odd + odd = even number of factors. If q has an even number of factors, then q^2 has an even + even = even number of factors. The same argument holds for p.) The prime factorization of $2q^2$ thus has an odd number of factors (an even number of factors for q^2 plus the factor of 2). This is a contradiction, as every number has a unique prime factorization, and $2q^2 = p^2$. Therefore, $\sqrt{2}$ must be irrational.

Technology

Connected Mathematics™ was developed with the belief that calculators should always be available and that students should decide when to use them. Students will need access to calculators for much of their work in this unit. Graphing calculators are required for the majority of the grade 8 Connected Mathematics units, so they are listed in this unit as well, but standard calculators that can calculate squares and square roots will suffice.

Mathematical and Problem-Solving Goals

Looking for *Pythagoras* was created to help students

- Make connections among coordinates, slope, distance, and area
- Relate the area of a square to the length of a side
- Develop strategies for finding the distance between two dots on a dot grid or two points on a coordinate grid
- Discover and apply the Pythagorean Theorem
- Extend understanding of number systems to include irrational numbers
- Locate irrational numbers on a number line
- Represent fractions as decimals and decimals as fractions
- Determine whether the decimal representation for a fraction terminates or repeats
- Use slopes to solve interesting problems

The overall goal of the Connected Mathematics curriculum is to help students develop sound mathematical habits. Through their work in this and other geometry units, students learn important questions to ask themselves about any situation that involves the principles explored in this unit, such as: *What is the length of a side of a square of a certain area? What is the relationship among the lengths of the sides of a right triangle? How can the Pythagorean Theorem be used to solve problems? How can knowing the slope of a line help to solve problems?*

Summary of Investigations

Investigation 1: Locating Points

Students review the concept of a coordinate grid and are introduced to finding the distance between pairs of points on a grid, setting the stage for developing strategies to find such distances. They also investigate geometric figures and their properties: given two vertices, they find other vertices that would define a square, a nonsquare rectangle, a right triangle, and a nonrectangular parallelogram.

Investigation 2: Finding Areas and Lengths

Students find areas of figures drawn on a dot grid and then explore the relationship between the area of a square and the length of its side. They are introduced to the concept of square root, and they develop a strategy for finding the distance between two points by analyzing the line segment between them: they draw a square using the segment as one side, find the area of the square, and then find the positive square root of the area. These lengths turn out to be either whole numbers or irrational numbers, such as $\sqrt{2}$.

Investigation 3: The Pythagorean Theorem

Students discover the Pythagorean Theorem and explore its implications. They collect information about the areas of the squares on the sides of right triangles and conjecture that the sum of the areas of the two smaller squares equals the area of the largest square, and they investigate a puzzle that verifies this conjecture. They apply the theorem to find the distance between two dots on a dot grid. Then, they apply the converse of the theorem to determine whether a triangle is a right triangle.

Investigation 4: Using the Pythagorean Theorem

Students use the Pythagorean Theorem to explore a variety of applications. They find distances on a baseball diamond; investigate the properties of some special right triangles, including a 30-60-90 triangle and an isosceles right triangle; and find missing lengths and angles in a group of triangles.

Investigation 5: Irrational Numbers

Students take a closer look at square roots. They express lengths as decimals, which leads to a study of decimal representations of fractions. They write fractions as terminating or repeating decimals and find fraction equivalents for terminating and repeating decimals. They are introduced to the concepts of rational numbers (those that can be represented by either terminating or repeating decimals) and irrational numbers (those that can be represented by nonterminating, nonrepeating decimals).

Investigation 6: Rational and Irrational Slopes

Students explore an interesting application of irrational numbers in the context of a video game in which the main character is trying to escape from a forest of trees planted in rows. Students review how to find the slope of a line and the connection between slope and points on a line and relate these concepts to rational and irrational numbers. They discover that, for a line not to pass through any grid point on an infinite grid, the line must have an irrational slope. The relationships between parallel lines (which have the same slope) and perpendicular lines (whose slopes are negative reciprocals) are explored in the ACE questions.

Connections to Other Units

The ideas in *Looking for Pythagoras* build on and connect to several big ideas in other Connected Mathematics units.

Big Idea	Prior Work	Future Work
calculating the distance between two points in the plane	measuring lengths *(Shapes and Designs; Covering and Surrounding)*; working with coordinates *(Variables and Patterns; Moving Straight Ahead; Thinking with Mathematical Models)*	finding midpoints of line segments *(Hubcaps, Kaleidoscopes, and Mirrors)*
finding areas of figures drawn on a coordinate grid with whole-number vertices	measuring areas of polygons and irregular figures *(Covering and Surrounding; Bits and Pieces I)*	studying transformations and symmetries of plane figures *(Hubcaps, Kaleidoscopes, and Mirrors)*
understanding square roots as lengths of sides of squares	applying the formula for area of a square *(Covering and Surrounding)*	looking for patterns in square numbers *(Frogs, Fleas, and Painted Cubes)*
understanding the Pythagorean Theorem and how it relates the areas of the squares on the sides of a right triangle	formulating, reading, and interpreting symbolic rules *(Variables and Patterns; Moving Straight Ahead; Thinking with Mathematical Models)*; working with the triangle inequality *(Shapes and Designs)*	formulating and using symbolic rules and the syntax for manipulating symbols *(Frogs, Fleas, and Painted Cubes; Say It with Symbols)*
using the Pythagorean Theorem to solve problems	solving problems in geometric and algebraic contexts *(Shapes and Designs; Moving Straight Ahead; Thinking with Mathematical Models)*	solving geometric and algebraic problems *(Frogs, Fleas, and Painted Cubes; Say It with Symbols; Hubcaps, Kaleidoscopes, and Mirrors)*
investigating rational numbers written as decimals	understanding fractions and decimals *(Bits and Pieces I; Bits and Pieces II)*	exploring sampling and approximations *(Samples and Populations)*
understanding irrational numbers as nonterminating, nonrepeating decimals	representing fractions as decimals and decimals as fractions *(Bits and Pieces I; Bits and Pieces II)*	solving quadratic equations *(Frogs, Fleas, and Painted Cubes; Say It with Symbols)*
understanding slope relationships of perpendicular and parallel lines	finding slopes of lines and investigating parallel lines *(Variables and Patterns; Moving Straight Ahead)*	investigating symmetry *(Hubcaps, Kaleidoscopes, and Mirrors)*

Materials

For students

- Labsheets
- Graphing calculators
- Geoboards (optional)
- Angle rulers (optional; transparent tools for measuring angles)
- Dot paper (provided as a blackline master)
- Centimeter grid paper (provided as a blackline master)
- String or straws
- Centimeter rulers (1 per student)
- Scissors

For the teacher

- Transparencies and transparency markers (optional)
- Transparencies of Labsheets 1.3, 2.1, 3.2, 4.2, and 5.1 (optional)
- Transparent dot paper (optional; copy the blackline master onto transparency film)
- Transparent centimeter grid (optional; copy the blackline master onto transparency film)
- Overhead geoboard (optional)

Resources

For students

Kolpas, Sidney J. *The Pythagorean Theorem: Eight Classic Proofs.* Palo Alto, Calif.: Dale Seymour Publications, 1992.

Pacing Chart

This pacing chart gives estimates of the class time required for each investigation and assessment piece. Shaded rows indicate opportunities for assessment.

Investigations and Assessments	Class Time
1 Locating Points	4 days
2 Finding Areas and Lengths	4 days
Check-Up	$\frac{1}{2}$ day
3 The Pythagorean Theorem	5 days
4 Using the Pythagorean Theorem	4 days
Quiz	1 day
5 Irrational Numbers	4 days
6 Rational and Irrational Slopes	3 days
Self-Assessment	Take home
Unit Test	1 day

Looking for Pythagoras Vocabulary

The following words and concepts are used in *Looking for Pythagoras.* Concepts in the left column are those essential for student understanding of this and future units. The Descriptive Glossary gives descriptions of many of these terms.

Essential terms developed in this unit
hypotenuse
irrational number
perpendicular
Pythagorean Theorem
rational number
real numbers
repeating decimal
square root
terminating decimal

Terms developed in previous units
angle
area
coordinate pair
coordinate system
coordinates
diagonal
equilateral triangle
horizontal
isosceles triangle
origin
parallel lines
parallelogram
perimeter
perpendicular lines
polygon
quadrant
quadrilateral
ratio
rectangle
right triangle
slope
square
square number
square unit
vertex, vertices
vertical
x-axis, y-axis

Nonessential terms
leg (of a right triangle)
midpoint
theorem

Assessment Summary

Embedded Assessment

Opportunities for informal assessment of student progress are embedded throughout *Looking for Pythagoras* in the problems, the ACE questions, and the Mathematical Reflections. Suggestions for observing as students explore and discover mathematical ideas, for probing to guide their progress in developing concepts and skills, and for questioning to determine their level of understanding can be found in the Launch, Explore, and Summarize sections of all investigation problems. Some examples:

- Investigation 2, Problem 2.2 *Launch* (page 26c) suggests questions you can use to assess what your students recall about the properties of equilateral triangles.
- Investigation 1, Problem 1.3 *Explore* (page 16e) suggests questions you can ask to help your students express how they are reasoning about the construction of geometric figures.
- Investigation 3, Problem 3.1 *Summarize* (page 40b) suggests questions you can ask to help your students explore some of the implications of the Pythagorean Theorem.

ACE Assignments

An ACE (Applications—Connections—Extensions) section appears at the end of each investigation. To help you assign ACE questions, a list of assignment choices is given in the margin next to the reduced student page for each problem. Each list indicates the ACE questions that students should be able to answer after they complete the problem.

Check-Up

One check-up, which may be given after Investigation 2, is provided for use as a quick quiz or warm-up activity. The check-up is designed for students to complete individually. You will find the check-up and its answer key in the Assessment Resources section.

Partner Quiz

One quiz, which may be given after Investigation 4, is provided with this unit. The quiz is designed to be completed by pairs of students with the opportunity for revision based on teacher feedback. You will find the quiz and its answer key in the Assessment Resources section. As an alternative to the quiz provided, you can construct your own quiz by combining questions from the Question Bank, this quiz, and unassigned ACE questions.

Question Bank

A Question Bank provides questions you can use for homework, reviews, or quizzes. You will find the Question Bank and its answer key in the Assessment Resources section.

Notebook/Journal

Students should have notebooks to record and organize their work. Notebooks should include student journals and sections for vocabulary, homework, quizzes, and check-ups. In their journals, students can take notes, solve investigation problems, and record their ideas about Mathematical Reflections questions. Journals should be assessed for completeness rather than correctness; they should be seen as "safe" places where students can try out their thinking.

A Notebook Checklist and a Self-Assessment are provided in the Assessment Resources section. The Notebook Checklist helps students organize their notebooks. The Self-Assessment guides students as they review their notebooks to determine which ideas they have mastered and which they still need to work on.

The Unit Test

The final assessment in *Looking for Pythagoras* is a unit test that focuses on applying the Pythagorean Theorem and the concepts of rational and irrational numbers.

Introducing Your Students to *Looking for Pythagoras*

Introduce this unit by asking students how they would find certain distances. "How would you calculate the distance between the classroom and the cafeteria?" "What is the distance across the classroom?" Their ideas about the first question will likely involve some sort of estimation strategy. For the second question, they may suggest pacing off the distance across the room, measuring with a meterstick, or counting ceiling or floor tiles.

Draw a square on a dot grid as shown below.

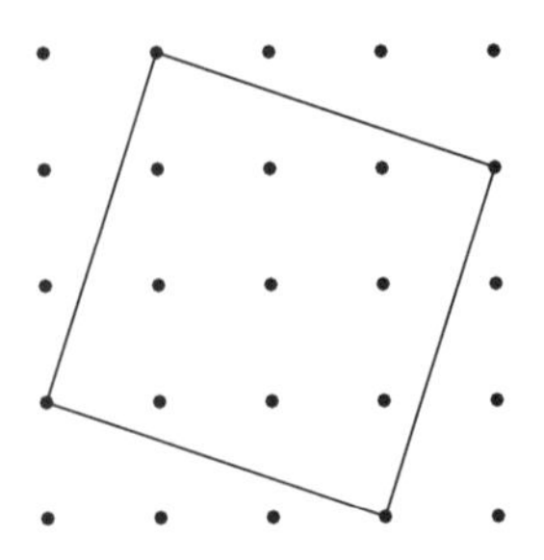

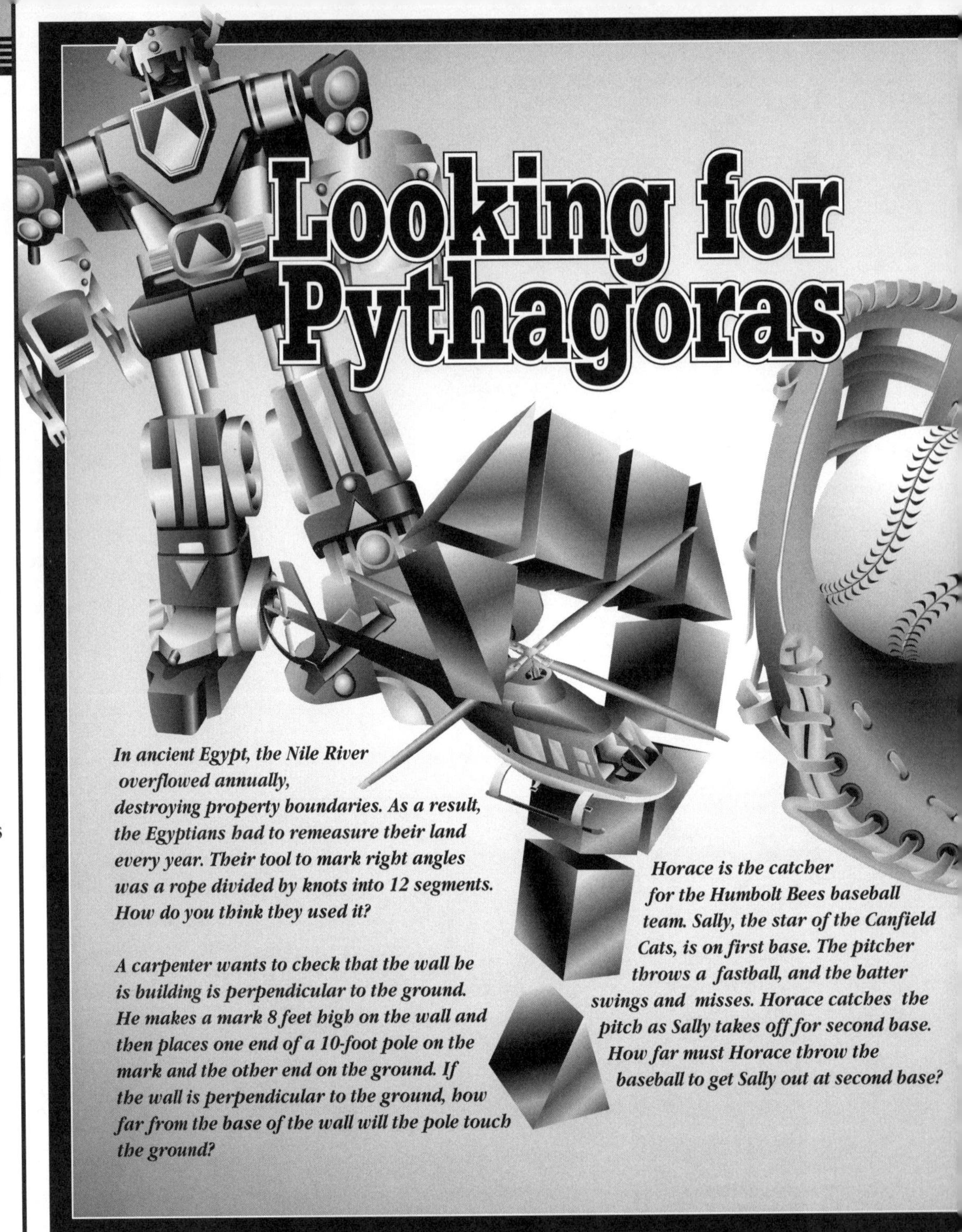

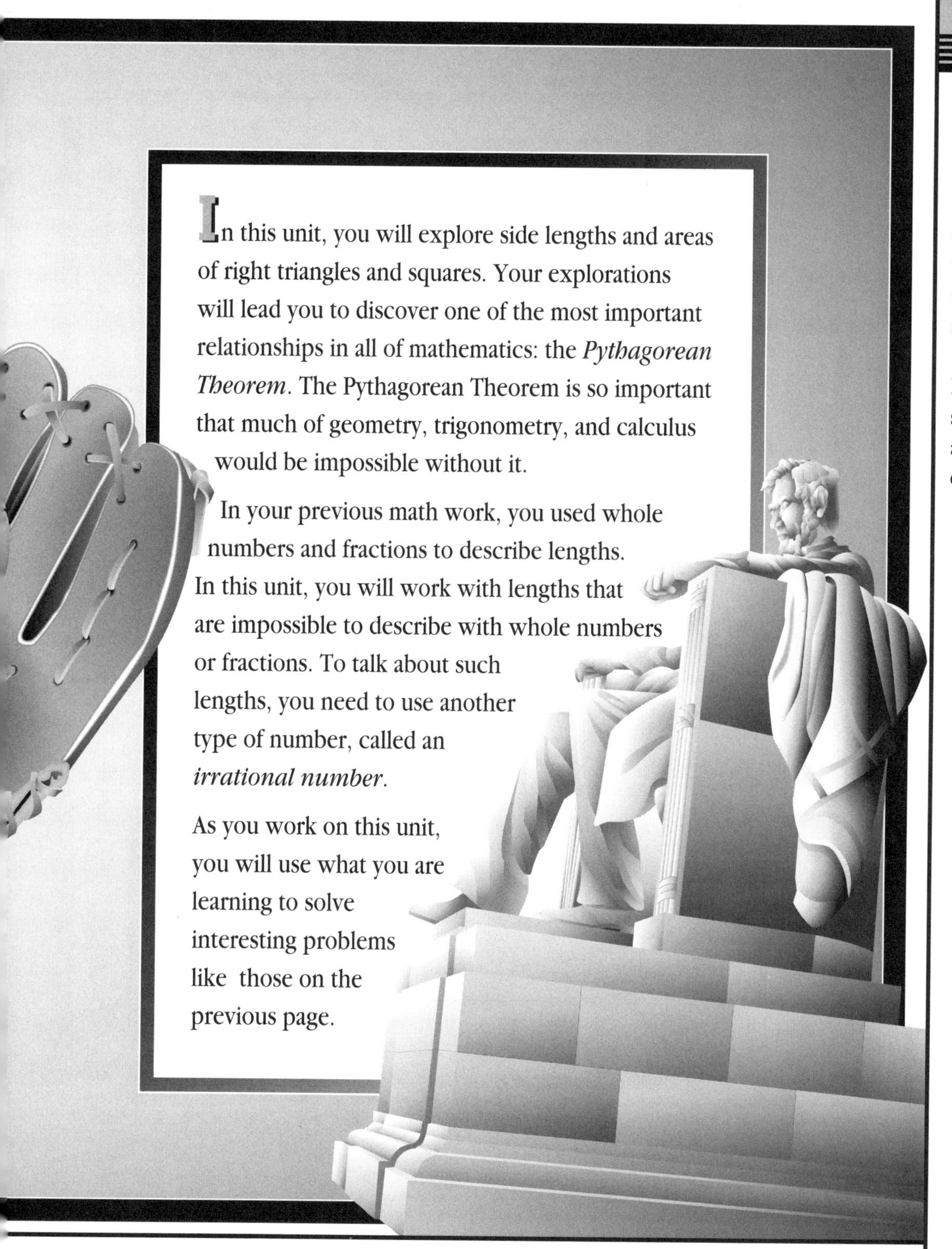

In this unit, you will explore side lengths and areas of right triangles and squares. Your explorations will lead you to discover one of the most important relationships in all of mathematics: the *Pythagorean Theorem*. The Pythagorean Theorem is so important that much of geometry, trigonometry, and calculus would be impossible without it.

In your previous math work, you used whole numbers and fractions to describe lengths. In this unit, you will work with lengths that are impossible to describe with whole numbers or fractions. To talk about such lengths, you need to use another type of number, called an *irrational number*.

As you work on this unit, you will use what you are learning to solve interesting problems like those on the previous page.

Let the class briefly discuss these questions: "What is the area of this figure?" "What is the length of each side of this figure?" "Is this figure a square? Why or why not?"

Explain that in this unit, students will explore the answers to these and related questions.

Mathematical Highlights

The Mathematical Highlights page provides information for students and for parents and other family members. It gives students a preview of the activities and problems in *Looking for Pythagoras.* As they work through the unit, students can refer back to the Mathematical Highlights page to review what they have learned and to preview what is still to come. This page also tells students' families what mathematical ideas and activities will be covered as the class works through *Looking for Pythagoras.*

Mathematical Highlights

In this unit, you will learn about a new type of number, and you will discover an important relationship among the side lengths of a right triangle.

- Working with figures drawn on a dot grid helps you develop efficient strategies for calculating areas.
- Finding the side lengths of squares drawn on a dot grid introduces you to the idea of square roots.
- Drawing a square and finding its area gives you a way to find the distance between two dots on a grid.
- Drawing squares on the sides of right triangles leads you to discover the Pythagorean Theorem.
- Applying the Pythagorean Theorem helps you solve interesting problems involving distances and lengths.
- As you investigate the side lengths of squares, you discover irrational numbers.
- Looking at patterns in the decimal representations of numbers helps you understand irrational numbers.
- By finding lines with slopes that are irrational numbers, you help a video game character escape from a laser forest.

Using a Calculator

In this unit, you will use a calculator to explore square roots and irrational numbers. As you work on the Connected Mathematics units, you decide whether to use a calculator to help you solve a problem.

The Investigations

The teaching materials for each investigation consist of three parts: an overview, student pages with teaching outlines, and detailed notes for teaching the investigation.

The overview of each investigation includes brief descriptions of the problems, the mathematical and problem-solving goals of the investigation, and a list of necessary materials.

Essential information for teaching the investigation is provided in the margins around the student pages. The "At a Glance" overviews are brief outlines of the Launch, Explore, and Summarize phases of each problem for reference as you work with the class. To help you assign homework, a list of "Assignment Choices" is provided next to each problem. Where space permits, answers to problems, follow-ups, ACE questions, and Mathematical Reflections appear next to the appropriate student pages.

The Teaching the Investigation section follows the student pages and is the heart of the Connected Mathematics curriculum. This section describes in detail the Launch, Explore, and Summarize phases for each problem. It includes all the information needed for teaching, along with suggestions for what you might say at key points in the teaching. Use this section to prepare lessons and as a guide for teaching the investigations.

Assessment Resources

The Assessment Resources section contains blackline masters and answer keys for the check-up, the quiz, the Question Bank, and the Unit Test. Blackline masters for the Notebook Checklist and the Self-Assessment are given. These instruments support student self-evaluation, an important aspect of assessment in the Connected Mathematics curriculum. A sample of one student's response to the Self-Assessment is included, along with a teacher's evaluation.

Blackline Masters

The Blackline Masters section includes masters for all labsheets and transparencies. Blackline masters of dot paper and centimeter grid paper are also provided.

Additional Practice

Practice pages for each investigation offer additional problems for students who need more practice with the basic concepts developed in the investigations as well as some continual review of earlier concepts.

Descriptive Glossary

The glossary provides descriptions and examples of the key concepts in *Looking for Pythagoras.* These descriptions are not intended to be formal definitions but are meant to give you an idea of how students might make sense of these important concepts.

Locating Points

In their work in this investigation, students review the concept of a coordinate grid and are introduced to finding the distance between two points on a grid.

In Problem 1.1, Driving Around Euclid, students analyze a map of the fictitious city of Euclid, in which streets are laid out on a coordinate grid. They find the distance a car would travel from one location to another, making the connection between the coordinates of two points and the distances between them. In Problem 1.2, Planning Emergency Routes, students compare the distances traveled by a car (which must travel along horizontal and vertical streets) and a helicopter (which takes the shortest, straight-line route) between two points. They use a ruler to find the helicopter distance. This problem sets the stage for developing strategies for finding the distance between two points on a grid. In Problem 1.3, Planning Parks, students investigate geometric figures. Given two vertices, they find other vertices that define a square, a nonsquare rectangle, a right triangle, and a nonrectangular parallelogram.

Mathematical and Problem-Solving Goals

- ***To review the use of coordinates for specifying locations***
- ***To use coordinates to specify direction and distance***
- ***To connect properties of geometric shapes, such as parallel sides, to coordinate representations***

Materials

Problem	For students	For the teacher
All	Graphing calculators	Transparencies: 1.1A to 1.3 (optional)
1.2	Centimeter rulers (1 per student)	Transparent centimeter ruler (optional)
1.3	Labsheet 1.3 (1 per student), centimeter grid paper, centimeter rulers (optional)	Transparency of Labsheet 1.3 (optional), transparent grid (optional)
ACE	Grid paper	

Locating Points

The map on the next page shows the central part of Washington, D.C. The streets form a grid of north-south and east-west lines. The city is divided into four sections, or *quadrants,* by the north-south line formed by North and South Capitol Streets and the east-west line stretching from the Lincoln Memorial, through the Mall, and down East Capitol Street.

U.S. Supreme Court Building

Think about this!

- How would you describe the locations of George Washington University, Dupont Circle, the White House, Union Station, and Benjamin Banneker Park?
- How would you describe the directions of Massachusetts Avenue, New York Avenue, and Vermont Avenue?
- How would you find the distance from Union Station to Dupont Circle?
- How do the locations of the intersection of G Street and 8th Street SE and the intersection of G Street and 8th Street NW relate to the location of the Capitol Building?

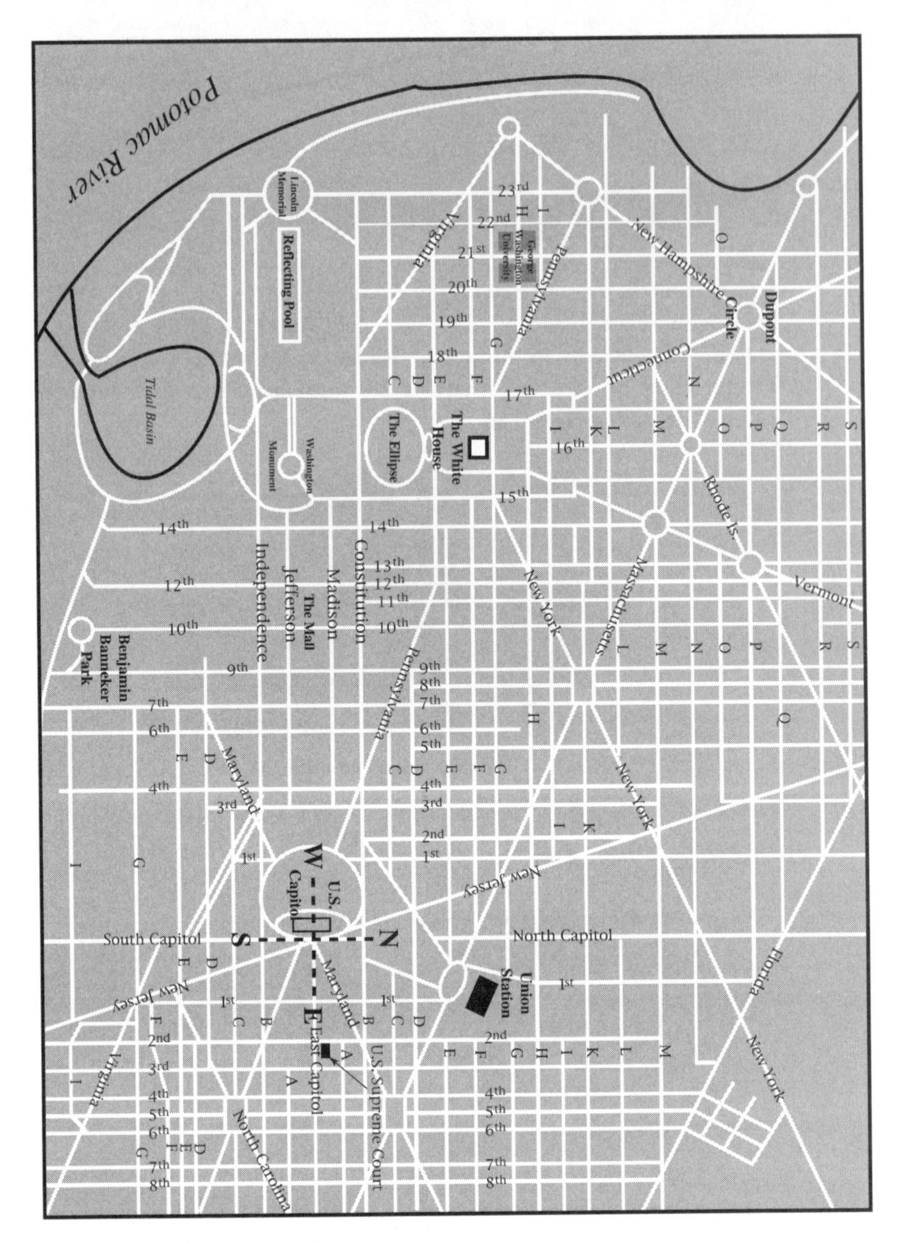

The street system of Washington, D.C., was designed 200 years ago by Pierre L'Enfant. L'Enfant's design is based on a *coordinate system,* with the origin at the Capitol and north-south and east-west streets forming grid lines.

In mathematics, we use a *coordinate system* to describe the locations of points. Recall that horizontal and vertical number lines, called the x- and y-axes, divide the plane into four quadrants. You describe the location of a point by giving its coordinates as an ordered pair of the form (x, y). On the coordinate grid below, four points are labeled with their coordinates.

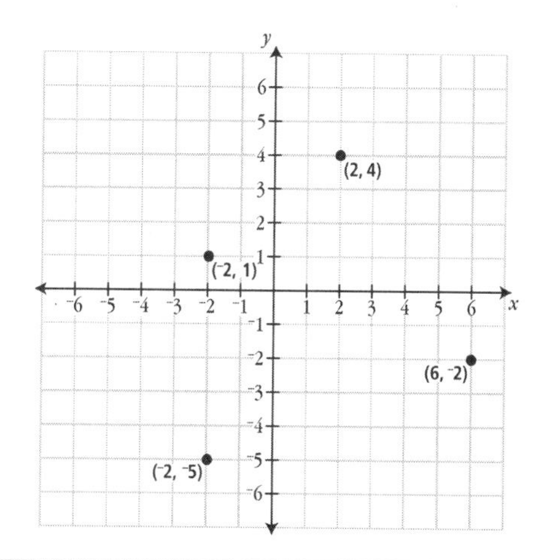

Did you know?

The coordinate system described above is called the *rectangular coordinate system.* You can also locate points in a plane by using the *polar coordinate system.* The polar coordinates of a point consist of a distance and an angle. Below is a polar grid with four points labeled with their coordinates. Can you figure out how to locate points on a polar grid?

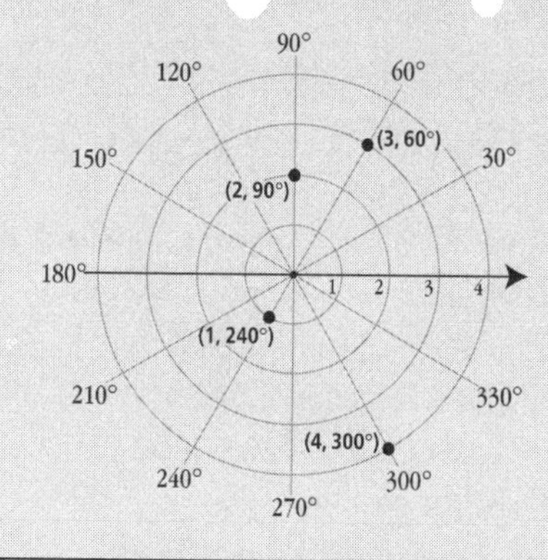

Driving Around Euclid

At a Glance

Grouping: *pairs*

Launch

- Talk about the Washington, D.C., map and how to locate landmarks.
- Introduce the map of Euclid.
- Have pairs work on the problem and follow-up.

Explore

- Help students connect the coordinates of two points to the distance by car between the points.

Summarize

- Review how the grid makes locating points easy.
- Talk about finding the distance by car between two points given their coordinates.
- Discuss the idea that distance and direction from the origin are needed to locate a point.

1.1 Driving Around Euclid

The founders of the city of Euclid liked math so much that they named their city after a famous mathematician and designed their street system to look like a coordinate grid. A map of the city is shown below. The Euclideans describe the locations of buildings and other landmarks by giving coordinates. For example, the animal shelter is located at (6, ⁻2).

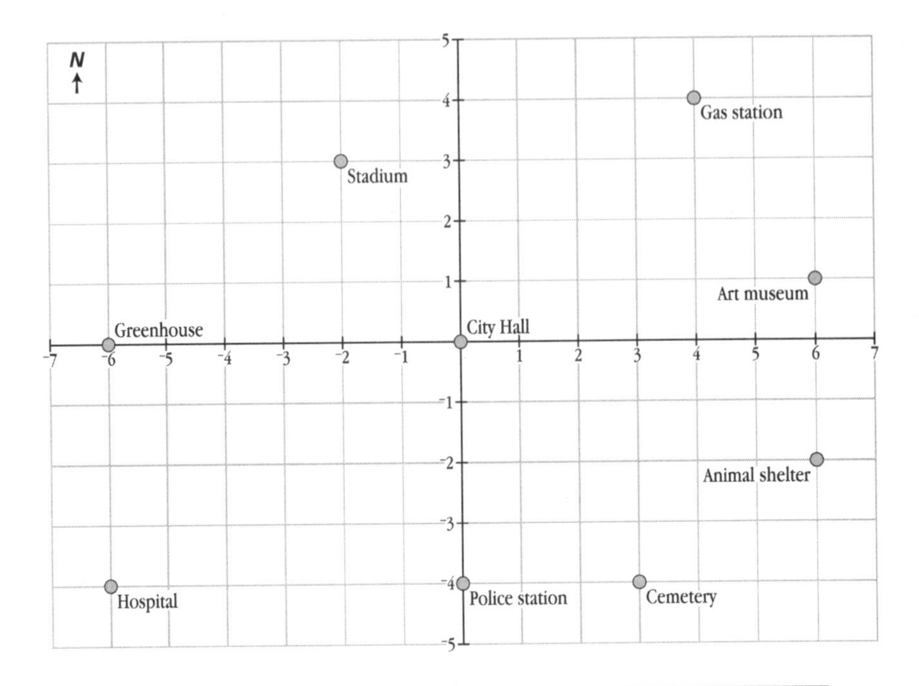

Problem 1.1

A. Give the coordinates of each labeled landmark on the map.

B. **1.** How many blocks would a car have to travel to get from the hospital to the cemetery?

2. How many blocks would a car have to travel to get from City Hall to the police station?

3. How many blocks would a car have to travel to get from the art museum to the gas station?

C. How can you tell the distance in blocks between two points if you know the coordinates of the points?

Assignment Choices

ACE questions 1, 3, 4, 7, and 8

Answers to Problem 1.1

A. City Hall (0, 0); greenhouse (⁻6, 0); cemetery (3, ⁻4); gas station (4, 4); hospital (⁻6, ⁻4); animal shelter (6, ⁻2); stadium (⁻2, 3); police station (0, ⁻4); art museum (6, 1)

B. 1. 9 blocks
2. 4 blocks
3. 5 blocks

C. Add the (positive) difference in the *x*-coordinates to the (positive) difference in the *y*-coordinates.

Problem 1.1 Follow-Up

1. Chloe is at City Hall. She asks a police officer where the Corner Ice-Cream Shop is, and he replies, "It's three blocks from here." Can Chloe find the ice-cream shop based on these directions? Where might the ice-cream shop be located?
2. The library is located at an intersection that is the same number of blocks from the stadium as the animal shelter is from the art museum. Where might the library be located?
3. Each street in Euclid is named with its equation. For example, the animal shelter is located on the corner of $X = 6$ Street and $Y = {}^{-}2$ Street. For each labeled landmark on the map, give the names of the two streets that intersect at that location.

1.2 Planning Emergency Routes

Euclid's chief of police has hired you to map out emergency routes that will allow paramedics and police officers to arrive at the scene of an accident as quickly as possible. She wants the plan to include routes for automobiles and helicopters.

Problem 1.2

The chief of police tells you that your plan must include the shortest routes between the following pairs of locations. Answer parts A and B for each pair of locations.

Pair 1: the police station to City Hall
Pair 2: the hospital to City Hall
Pair 3: the hospital to the art museum
Pair 4: the police station to the stadium

A. **1.** Give the coordinates of each location, and give precise directions for an emergency car route from the starting location to the ending location.

2. Find the total distance, in blocks, a police car would have to travel to get from the starting location to the ending location along your route.

B. A helicopter can travel directly from one point to another. Find the total distance, in blocks, a helicopter would have to travel to get from the starting location to the ending location. You may find it helpful to use a centimeter ruler (1 centimeter = 1 block).

1.2 Planning Emergency Routes

At a Glance

Grouping: ***individuals, then pairs***

Launch

- Talk about the emergency routes being planned for Euclid.
- Discuss using a ruler to measure distances by helicopter on the map.
- Have students explore the problem individually.

Explore

- When students finish the problem, have pairs of students share answers and then work on the follow-up.

Summarize

- Have the class verify answers to the problem and follow-up.
- Discuss the triangle inequality and its application in the follow-up.

Answers to Problem 1.1 Follow-Up

See page 16f.

Answers to Problem 1.2

A. 1. Possible answer: *Pair 1:* From the police station at $(0, {}^{-}4)$ to City Hall at $(0, 0)$, go north (up) 4 blocks. *Pair 2:* From the hospital at $({}^{-}6, {}^{-}4)$ to City Hall at $(0, 0)$, go north (up) 4 blocks and east (right) 6 blocks. *Pair 3:* From the hospital at $({}^{-}6, {}^{-}4)$ to the art museum at $(6, 1)$, go north (up) 5 blocks and east (right) 12 blocks. *Pair 4:* From the police station at $(0, {}^{-}4)$ to the stadium at $({}^{-}2, 3)$, go west (left) 2 blocks and north (up) 7 blocks.

2. *Pair 1:* 4 blocks; *Pair 2:* 10 blocks; *Pair 3:* 17 blocks; *Pair 4:* 9 blocks

B. *Pair 1:* 4 blocks; *Pair 2:* about 7.2 blocks; *Pair 3:* 13 blocks; *Pair 4:* about 7.3 blocks

Assignment Choices

ACE questions 2, 9–11, and unassigned choices from earlier problems

Planning Parks

At a Glance

Grouping:
small groups

Launch

- Talk about the plan for parks in Euclid.
- Discuss the special properties of a figure that makes it a square, a rectangle, a parallelogram, or a right triangle.
- Distribute Labsheet 1.3 and centimeter grid paper, and have groups of three or four work on the problem and follow-up.

Explore

- As students work, ask questions about their reasoning.

Summarize

- For each park shape, have students offer possible vertices and explain how they found them.
- Discuss students' conjectures about diagonals of quadrilaterals.

Problem 1.2 Follow-Up

1. How much farther is it from the greenhouse to the stadium by car than by helicopter?
2. How much farther is it from the hospital to the art museum by car than by helicopter?
3. Will the helicopter distance between two locations always be shorter than the car distance? Explain your reasoning.

1.3 Planning Parks

The Euclid city council is planning to develop parks with geometric shapes. For some of the parks, the city council has given constraints to the park designers. For example, one of the parks must have a border that includes the vertices (1, 1) and (4, 2).

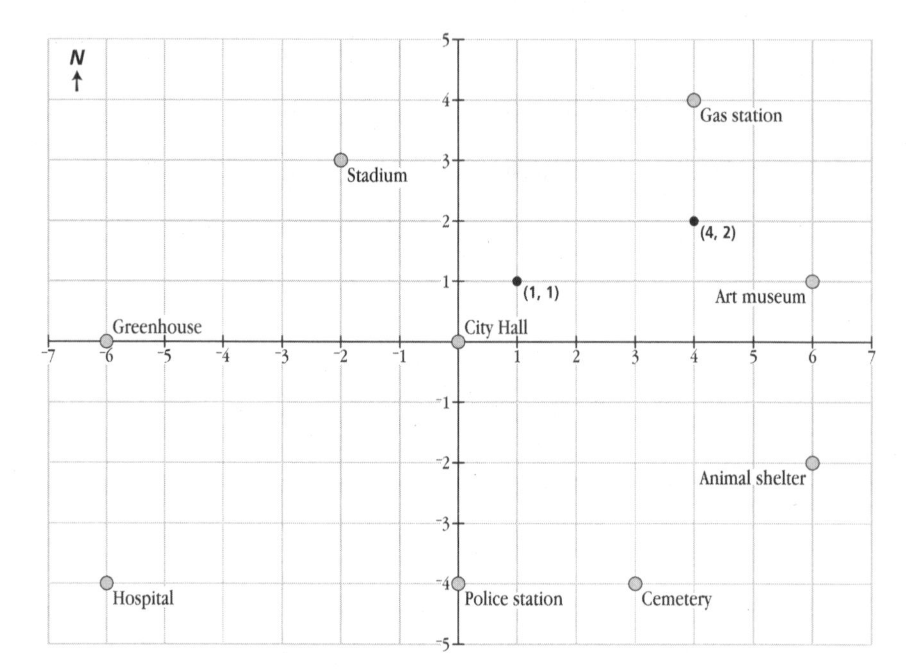

Assignment Choices

ACE questions 5, 6, 12, 13, and unassigned choices from earlier problems

Answers to Problem 1.2 Follow-Up

1. By car it is 7 blocks and by helicopter it is 5 blocks, so it is 2 blocks farther by car.
2. By car it is 17 blocks and by helicopter it is 13 blocks, so it is 4 blocks farther by car.
3. The helicopter distance will always be shorter than the car distance unless the points are on the same vertical or horizontal line, in which case the distances will be equal. If the points aren't on the same vertical or horizontal line, the car must travel farther than the helicopter because it can't travel in a straight line, which is the shortest distance between two points.

Problem 1.3

A. If the park with the given vertices is to be a square, what could the coordinates of the other two vertices be? Give two answers.

B. If this park is to be a nonsquare rectangle, what could the coordinates of the other two vertices be? Give two answers.

C. If this park is to be a right triangle, what could the coordinates of the other vertex be? Give two answers.

D. If this park is to be a parallelogram that is *not* a rectangle, what could the coordinates of the other two vertices be? Give two answers.

on geoboard

Problem 1.3 Follow-Up

Draw the diagonals of the quadrilaterals you found in Problem 1.3. For each quadrilateral, look at the lengths of the diagonals and the angles at which the diagonals intersect. Use your observations to make a conjecture about the diagonals of a square, a nonsquare rectangle, and a nonrectangular parallelogram. Test your conjecture on three different quadrilaterals.

Answers to Problem 1.3

See page 16g.

Answer to Problem 1.3 Follow-Up

Possible conjectures: The diagonals of any rectangle, including squares, are equal in length. The diagonals of nonrectangular parallelograms are not equal in length. The diagonals of a square intersect at right angles (or are perpendicular).

Answers

Applications

1. For routes that are 10 blocks long (the shortest distance), there are five possible halfway points: (⁻5, 0), (⁻4, ⁻1), (⁻3, ⁻2), (⁻2, –3), and (⁻1, ⁻4).

2. For the straight-line helicopter route, there is only one halfway point: (⁻3, ⁻2).

3a. the art museum and the cemetery

3b. Possible answer: To get to the art museum, drive 6 blocks east, turn left, and go north 1 block. To get to the cemetery, drive 3 blocks east, turn right, and drive 4 blocks south.

4. The hospital is 4 blocks from the greenhouse. There are ten intersections on the map that are 4 blocks by car from the gas station: (1, 5), (0, 4), (1, 3), (2, 2), (3, 1), (4, 0), (5, 1), (6, 2), (7, 3), and (7, 5).

Applications • Connections • Extensions

As you work on these ACE questions, use your calculator whenever you need it.

Applications

In 1–4, use the map of Euclid on page 8.

1. What are the coordinates of a point halfway from City Hall to the hospital if you are traveling by taxi? Is there more than one answer to this question? Explain.

2. What are the coordinates of a point halfway from City Hall to the hospital if you are traveling by helicopter? Is there more than one answer to this question? Explain.

3. **a.** Which landmarks on the map are seven blocks from City Hall by car?

b. For each landmark you listed in part a, what information could you give someone so he or she would know exactly how to drive from City Hall to the landmark?

4. Euclid Middle School is located at an intersection on the map that is the same distance by car from the gas station as the hospital is from the greenhouse. List the coordinates of each place where the school might be located.

5. The points (0, 0) and (3, 2) are two vertices of a polygon.

   a. If the polygon is a square, what might the other two vertices be? Give two answers.

   b. If the polygon is a nonrectangular parallelogram, what might the other two vertices be? Give two answers.

   c. If the polygon is a right triangle, what might the other vertex be? Give two answers.

6. The points (3, 3) and (2, 6) are two vertices of a right triangle.

   a. List at least three points that could be the third vertex.

   b. How many right triangles with vertices (3, 3) and (2, 6) can you draw? Explain.

Connections

In 7 and 8, use the map of Euclid on page 8.

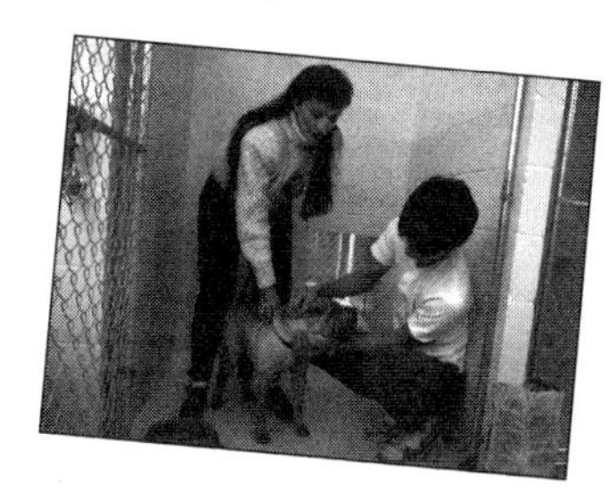

7. In the city of Euclid, a block is 150 meters long.

   a. How many meters is it from City Hall to the animal shelter if you travel by car?

   b. How many meters is it from the police station to the gas station if you travel by car?

8. a. Ki-Cho walked two blocks west from the Euclid police station and then walked three blocks north. Give the coordinates of the place where he stopped.

   b. Plimpton left the Euclid stadium and walked three blocks east, three blocks south, two blocks west, and four blocks north. Give the coordinates of the place where he stopped.

5a. (−2, 3) and (1, 5); (5, −1) and (2, −3)

5b. There is an infinite number of pairs possible, including (2, 0) and (5, 2); (0, 2) and (3, 4); (0, −2) and (3, 0); and (2, −1) and (5, 1).

5c. There is an infinite number of possible vertices, including (0, 2), (3, 0), (4, −6), (5, −1), (−2, 3), and (1, 5).

6a. Possible answer: (2, 3), (3, 6), and (−3, 1)

6b. An infinite number of right triangles can be drawn. The third vertex can be located at any point on the line that goes through (0, 2) and (6, 4) (the line $y = \frac{1}{3}x + 2$) or on the line that goes through (−1, 5) and (5, 7) (the line $y = \frac{1}{3}x + \frac{16}{3}$). Each of these lines is perpendicular to the segment connecting (3, 3) and (2, 6), so these lines create the right angle for the triangle. Some students may express this idea as follows: Imagine a line starting from one of the given points and at a right angle to the given side. Any point along that line can be the third vertex of the triangle.

Connections

7a. 8 blocks × 150 meters/block = 1200 meters

7b. 12 blocks × 150 meters/block = 1800 meters

8a. (−2, −1)

8b. (−1, 4)

8c. There are three possible shortest routes. For example, Cassandra could walk 2 blocks west and 1 block south.

8d. There are five possible shortest routes. For example, Aida could walk 1 block west and 4 blocks north.

8e. Figure out how many blocks east or west you have to go and how many blocks north or south you have to go. The total is the number of blocks in the shortest route. Any route that involves going east or west that number of blocks and going north or south that number of blocks will be a shortest route.

9a. **i.** (1, F)

ii. (16, F)

iii. (10, I)

9b. No, a coordinate pair does not determine a unique point. To describe a unique point, you could use a negative sign to indicate a coordinate west or south of the Capitol Building.

10. Road maps are typically partitioned into square areas by consecutive letters running along the sides of the map and consecutive numbers running along the top and bottom. This system is similar to a coordinate grid system, but the letters and numbers do not refer to points; they refer to regions. For example, anything in the top-left square might be referred to as located in region A-1.

c. Ki-Cho's friend Cassandra just left her job at City Hall. She is supposed to meet Ki-Cho at his ending location from part a. If she walks along city streets, what is the shortest route she could take? Is there more than one possible route?

d. Plimpton's sister Aida wants to meet him at his ending location from part b. She is now at City Hall. If she walks along city streets, what is the shortest route she could take? Is there more than one possible route?

e. In general, how can you find the shortest route from City Hall to any point in Euclid?

9. Look at the map of Washington, D.C., on page 6. You can think of the map as a coordinate grid with the origin at the Capitol Building. The coordinates of an intersection on the map would be in the form (vertical street name, horizontal street name). For example, the coordinates of George Washington University would be (21, H).

a. Give the coordinates of each landmark.

i. Union Station **ii.** White House **iii.** Benjamin Banneker Park

b. Does a coordinate pair determine a unique point on the map? If not, what additional information could you give to describe a unique point?

10. Find a road map of your state. Figure out how to use the map's index to locate a city, street, or other landmark. How is finding a landmark on a road map by using its index description similar to and different from finding a landmark in Euclid by using its coordinates?

11. Use a map of your state to plan an airplane trip from your city or town to four other locations in your state. Write a set of directions for your trip that you could give to the pilot.

11. Answers will vary. Students should include compass directions as well as distances and will need to decide where the distances are to be measured from, such as airports or city centers. For example: *Starting at the airport at Grand Rapids, go south 47 miles to the airport at Kalamazoo. From Kalamazoo, go northeast 60 miles to the airport at Lansing. From Lansing, go southeast 80 miles to the airport at Detroit.* (**Teaching Tip:** You may want to point out that pilots need more exact directions than just saying "north" or the like, because the actual direction may be a few degrees east or west of due north.)

Extensions

12. *Perpendicular lines* intersect at right angles. On grid paper, draw several parallelograms with diagonals that are perpendicular. What do you observe about these parallelograms?

13. In the LOGO programming language, the computer screen is considered to be a coordinate grid with the point (0, 0) at the starting, or home, location of the turtle. Write a set of LOGO commands for this delivery route in the city of Euclid: from City Hall to the greenhouse, to the cemetery, to the gas station, to the stadium, and back to City Hall. Assume that each block is 25 turtle steps.

Review of LOGO commands

`fd` The *forward* command tells the turtle to move forward. The `fd` command must be followed by a space and a number telling the turtle how many steps to take. For example, `fd 50` tells the turtle to move forward 50 steps.

`lt` The *left turn* command tells the turtle to turn to its left. The `lt` command must be followed by a space and a number telling the turtle how many degrees to turn. For example, `lt 90` tells the turtle to turn left 90°.

`rt` The *right turn* command tells the turtle to turn to its right. The `rt` command must be followed by a space and a number telling the turtle how many degrees to turn. For example, `rt 45` tells the turtle to turn right 45°.

Extensions

12. See below left.

13. Possible answer:

```
lt 90
fd 150
rt 180
fd 225
rt 90
fd 100
lt 180
fd 200
rt 90
fd 25
rt 180
fd 150
lt 90
fd 25
fd 75
lt 90
fd 50
```

12. Possible answer: A rhombus is the only parallelogram with perpendicular diagonals. (**Teaching Tip:** Students may only say that squares—rhombuses with right angles—have perpendicular diagonals; you may want to encourage them to look for nonsquare rhombuses.)

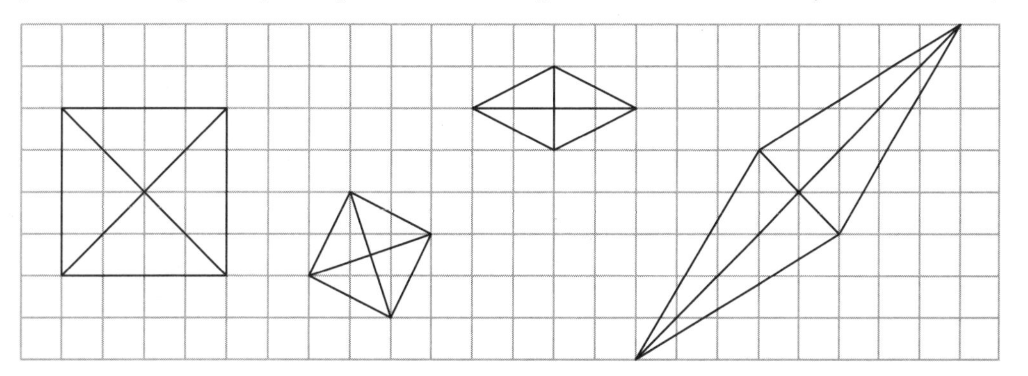

Possible Answers

1. Car distances are the same as or greater than helicopter distances. If the two places do not lie on the same vertical or horizontal line, the helicopter distance will be shorter because the car won't be able to travel in a straight line between them but the helicopter will.

2. The car distance between two landmarks is the sum of the positive differences of the *x*- and *y*-coordinates. (Or, the car distance is the sum of the absolute value of the differences between the *x*- and *y*-coordinates.) The helicopter distance can be measured with a ruler.

3. The points are the vertices of a parallelogram. Opposite sides have equal lengths and slope.

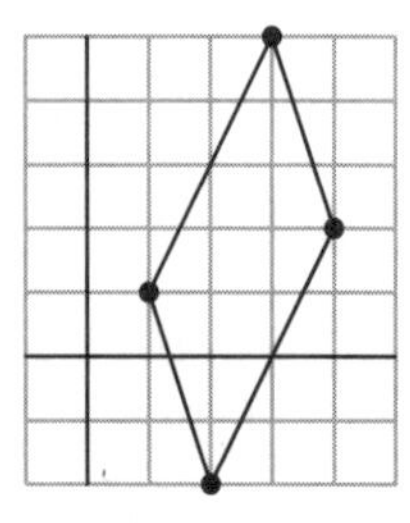

(Note: The slopes may be compared intuitively at this time. Students may say the distance between parallel lines is always the same, or they may use left/right, up/down language to express this idea. Others may find the actual slopes. This type of problem will be revisited at the end of the unit when students have more information on how to find lengths of segments and slopes of lines.)

Mathematical Reflections

In this investigation, you reviewed how to locate points and find distances on a coordinate grid. You also reviewed the geometric properties of some common polygons. These questions will help you summarize what you have learned:

1. In the city of Euclid, how does the distance a car needs to travel to get from one place to another compare to the distance a helicopter needs to travel?
2. If you know the coordinates of two landmarks in Euclid, how can you find the distance in blocks between them?
3. Can the following points be connected to form a parallelogram? Explain your answer.

   (1, 1) (2, ⁻2) (4, 2) (3, 5)

Think about your answers to these questions, discuss your ideas with other students and your teacher, and then write a summary of your findings in your journal.

Tips for the Linguistically Diverse Classroom

Diagram Code The Diagram Code technique is described in detail in *Getting to Know Connected Mathematics*. Students use a minimal number of words and drawings, diagrams, or symbols to respond to questions that require writing. Example: Question 1—A student might answer this question by drawing a car and a helicopter, writing *distance* below the car and *distance* below the helicopter, and putting ≥ between them to mean *car distance is greater than or equal to helicopter distance.* The student might also sketch an example of two places that do not lie on the same vertical or horizontal line, a helicopter, a straight line between the points, and label the drawing *shorter distance.*

TEACHING THE INVESTIGATION

1.1 • Driving Around Euclid

In this problem, students review the concept of the coordinate grid and are introduced to the idea of determining the distance between two points.

Launch

To introduce the topic, talk with the class about the map of Washington, D.C., shown in the student edition.

> How could you locate various places on this map in relation to the White House? Is that an efficient method?

Next, talk about the map of the fictitious city of Euclid, which is also shown on Transparency 1.1A. Draw attention to the origin (the location of City Hall) and the meaning of the coordinates. Help students to understand that a grid system is convenient for locating points on a map, but only if we know where to count from and what scale is being used.

> What are the coordinates of City Hall? *(0, 0)*
>
> What are the coordinates of the art museum? *(6, 1)* What do the 6 and the 1 mean? *(The numbers indicate that the art museum is 6 blocks to the right and 1 block up from the origin, or City Hall.)*
>
> Is there more than one way to travel from City Hall to the art museum? *(yes)*
>
> What is the shortest distance along the streets of Euclid from City Hall to the art museum? *(7 blocks)*
>
> Is there more than one shortest path from City Hall to the art museum? *(There are several, such as right 2 blocks, up 1 block, and right 4 blocks.)*

When students seem confident about reading map coordinates and finding distances, have them work in pairs on the problem and follow-up.

Explore

As students work, encourage them to look for connections between the coordinates of two points and the distance in blocks between the points, as traveled by a car.

> What information does the first number in each coordinate pair tell us about the distance between the two locations? What information does the second number tell us? What is the total distance?

Summarize

In the summary, establish that students understand that the grid system makes it possible to refer to each landmark in Euclid by a unique locator.

Why might it be important to be able to locate places in a city by using a simple system like grid coordinates?

What information do you need to be able to locate a point on a grid?

When we specify a point in Euclid, where are we counting from? What scale are we using? *(We count from City Hall, and the scale is in number of blocks.)*

Be sure students can interpret the x- and y-coordinates of a point. Given a point on the grid, they should be able to name the coordinates. Given the coordinates of a point, they should be able to locate the point on the grid.

Extend the coordinate idea to include non-integers.

Where in Euclid is the point $(2, 3\frac{1}{2})$?

Distance on the map is expressed in terms of the number of blocks a car would travel from one place to another. Part C helps students make the connection between the car distance and the absolute distance between two points on the map grid. Talk with the class about finding the distance between two points given their coordinates.

For example, the hospital and the cemetery are on the same horizontal line; City Hall and the police station are on the same vertical line. To find the distance between these pairs of points, you simply find the difference between the x-coordinates or y-coordinates, respectively; the difference between the other pairs of coordinates is 0.

The distance between the art museum and the gas station, which do not lie on the same vertical or horizontal line, is the sum of the difference between the x-coordinates and the difference between the y-coordinates. Actually, as distance is always positive, it is the sum of the ***absolute value*** of the difference between the x-coordinates and the ***absolute value*** of the difference between the y-coordinates. (The concept of using absolute value to express distance is explored in the grade 7 unit *Accentuate the Negative.*)

To go from the art museum to the gas station, how many blocks do you travel in a horizontal direction? *(2 blocks)* How is this distance related to the coordinates of the points? *(It is the positive difference, or the absolute value of the difference, between the x-coordinates.)*

To go from the art museum to the gas station, how many blocks do you travel in a vertical direction? *(3 blocks)* How is this distance related to the coordinates of the points? *(It is the positive difference, or the absolute value of the difference, between the y-coordinates.)*

Is there more than one path from the art museum to the gas station that is 5 blocks long? *(Yes, there are several.)*

Follow-up questions 1 and 2 reinforce the idea that both the distance and the direction from the origin must be indicated in order to locate a point. To review this concept, ask the class for more information that would precisely locate the ice-cream shop and the library.

> What information is needed to precisely locate the ice-cream shop? *(The direction of the shop from City Hall is needed; you could say, for example, that the shop is 3 blocks south or 3 blocks west of City Hall.)*
>
> What information is needed to precisely locate the library? *(Again, you need to know direction. The distance is 3 blocks, so to locate the library exactly, you could say, for example, that it is 3 blocks up from the stadium, or 2 blocks to the right and 1 block down from the stadium.)*

Verify that everyone understands that to precisely locate a position on the grid, a vertical distance and a horizontal distance and the direction of each must be given.

1.2 • Planning Emergency Routes

In this problem, students find the distance between two points in two ways: the distance along grid lines (represented by the north-south and east-west distances a car would travel) and the straight-line distance (represented by the distance a helicopter would travel).

Launch

Talk about the chief of police's concern about automobile and helicopter emergency routes in Euclid. Ask:

> How might you determine the distance a helicopter would travel to get from one point to another in Euclid?

If no one suggests using a ruler, explain that since each centimeter on the map represents one block, a centimeter ruler could be used to find the straight-line distance between two points.

Have students work on the problem individually.

Explore

As students work, check to see how they are finding the distance traveled by a helicopter between two points.

Have students meet in pairs to check answers. When pairs are ready, they can work on the follow-up.

Summarize

Have the class verify their answers to the problem and follow-up.

In the follow-up, students should have discovered the application of the concept that the shortest distance between two points is a straight line—in this case, the distance traveled by a helicopter. Review with the class why the helicopter distance is always shorter than or equal to the car distance. The triangle inequality, which students may have encountered in their earlier mathematics work, states that *the sum of the lengths of any two sides of a triangle is greater than the length of the third side.* The car distance is the sum of the lengths of two sides of a triangle; the helicopter distance is the length of the third side.

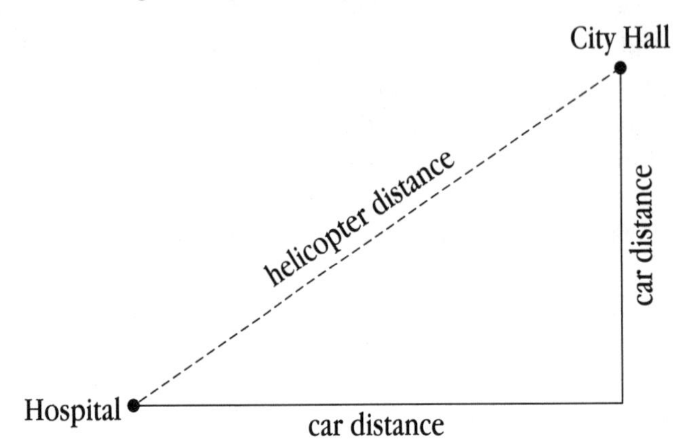

This is an opportunity to verify that students connect directions on a coordinate grid with map directions. Going left is traveling west, going up is traveling north, and so on.

1.3 • Planning Parks

In this problem, students investigate the properties of several geometric shapes. Given the coordinates of two vertices of a quadrilateral, they are asked to find other coordinates that will form a square, a nonsquare rectangle, a right triangle, and a nonrectangular parallelogram.

Launch

Tell the story about park planning in Euclid. Discuss describing the shapes of the parks by giving the vertices of their borders. Make sure students understand what defines a square, a right triangle, a rectangle, and a parallelogram. You may want to display a transparent grid, add a set of coordinate axes, and ask:

> Suppose a park with the two given vertices were made in the shape of a right triangle. Locate a right triangle on this grid, and tell us the coordinates of the third vertex. How do you know that this is a right triangle?
>
> Now locate a rectangle that has one of its vertices at the origin. Tell us the coordinates of its vertices. How do you know that this is a rectangle?

Take this opportunity to assess what students know about the properties of squares, rectangles, right triangles, and parallelograms. Do they know that squares have sides of equal length and four right angles? Do they know that a figure's orientation does not matter? Do they know that parallelograms have two pairs of parallel sides?

Now, describe Problem 1.3. Distribute Labsheet 1.3 (which contains two copies of the Euclid map) and centimeter grid paper, and have students work in groups of three or four on the problem and follow-up.

Explore

As students work, ask questions about the reasoning they are applying.

> How did you figure out where to put the vertices so that this park's sides would all be the same length?
>
> How did you determine where to put the vertices so that the sides of this park would be parallel?
>
> How did you decide where these vertices had to be to create these right angles?

Encourage students to discuss with the others in their group how they are finding the vertices of each shape so that each student can explain the group's strategies.

Summarize

Ask students to share their strategies for finding the vertices for each park shape. Here are some strategies that students might have used:

- Students might intuitively use the concept of slope to check that opposite sides of a quadrilateral are parallel.
- Students might recall (perhaps from the grade 7 unit *Moving Straight Ahead*) that parallel lines have the same slope and then use this fact to establish parallel lines.
- Students might apply the concept of slope by counting units up and units over to match the slope of an existing line segment.
- To check for right angles, students might use the corner of a piece of paper.
- Students might apply the fact that vertical and horizontal lines are perpendicular (though it is unlikely at this stage that they will know that the slopes of perpendicular lines are negative reciprocals).
- Students might use rulers or the marked edge of a piece of paper to check lengths.
- Students might use angle rulers to measure angles.
- To find a park in the shape of a right triangle, students might divide a rectangle or a square in half along one of its diagonals.

In parts A, B, and D, if no one offers a park shape in which a line segment through the given vertices forms a diagonal of the shape rather than a side, introduce this possibility.

Review the follow-up, taking time to gather students' observations about the diagonals in the shapes. Their conjectures about length or perpendicularity of the diagonals will be verified in Investigation 6. The diagonals of squares and rectangles have equal length. Some students may notice that the diagonals of a square are also perpendicular.

As a final summary, put a transparent grid on the overhead, and label x- and y-axes. Draw several parallelograms on the grid.

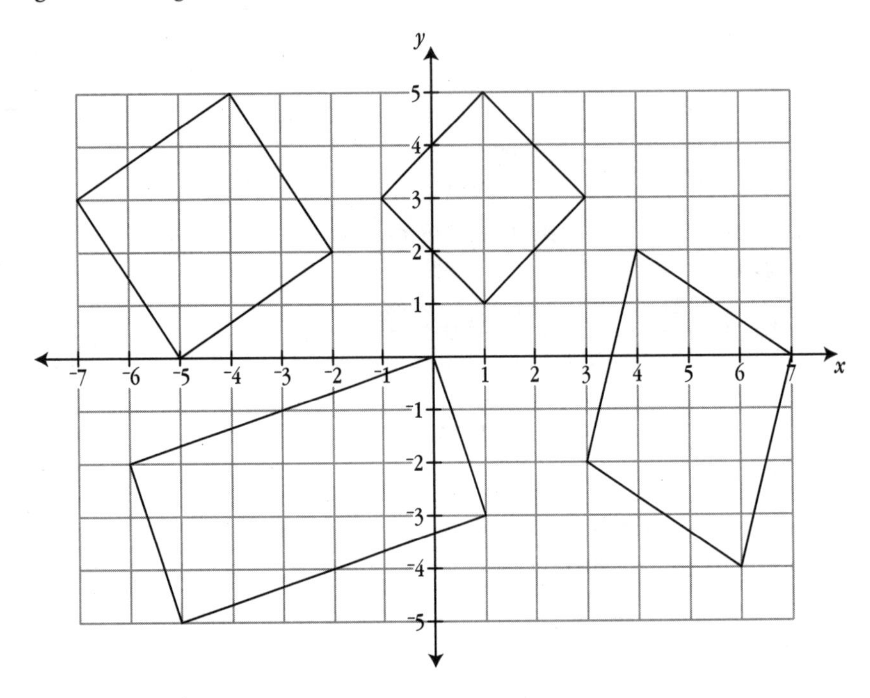

Ask the students to explain what is special about each figure—for example, a parallelogram may be a square, a rectangle but not a square, a parallelogram but not a rectangle, or a figure with four equal sides that is not a square.

Additional Answers

Answers to Problem 1.1 Follow-Up

1. Chloe would have to look in several places to find the ice-cream shop, as it could be at (3, 0), (0, 3), (⁻3, 0), (0, ⁻3), (1, 2), (1, ⁻2), (2, 1), (2, ⁻1), (⁻1, 2), (⁻1, ⁻2), (⁻2, 1), or (⁻2, ⁻1).
2. The animal shelter is three blocks from the art museum. Eleven intersections shown on the map are three blocks from the stadium: (⁻1, 5), (0, 4), (1, 3), (0, 2), (⁻1, 1), (⁻2, 0), (⁻3, 1), (⁻4, 2), (⁻5, 3), (⁻4, 4), and (⁻3, 5).
3. City Hall (X = 0 Street and Y = 0 Street)
   stadium (X = ⁻2 Street and Y = 3 Street)
   hospital (X = ⁻6 Street and Y = ⁻4 Street)
   cemetery (X = 3 Street and Y = ⁻4 Street)
   gas station (X = 4 Street and Y = 4 Street)
   greenhouse (X = ⁻6 Street and Y = 0 Street)
   police station (X = 0 Street and Y = ⁻4 Street)
   art museum (X = 6 Street and Y = 1 Street)

Answers to Problem 1.3

(Note: The answers given below use integer coordinates. Students might also give non-integer coordinates.)

A. There are three possible pairs of vertices: (3, 5) and (0, 4); (5, $^-$1) and (2, $^-$2); and (3, 0) and (2, 3).

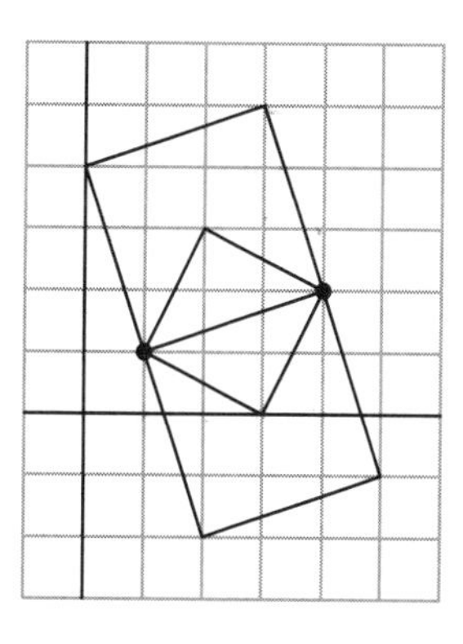

B. There are many possible pairs of vertices, such as (6, $^-$4) and (3, $^-$5); (1, 2) and (4, 1); and (2, 0) and (3, 3).

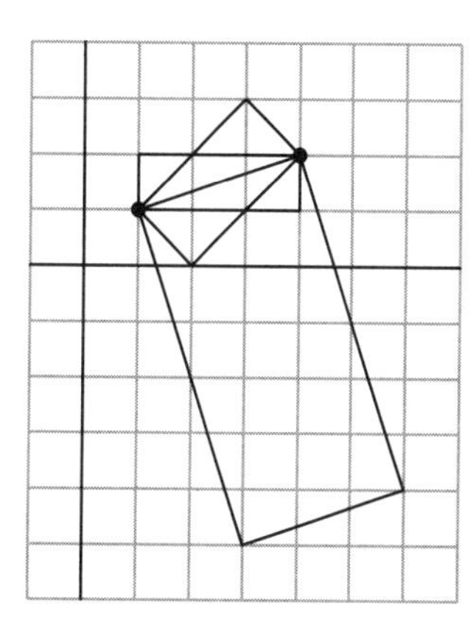

C. There are several possible vertices, such as (3, $^-$5), (2, 3), and (5, $^-$1).

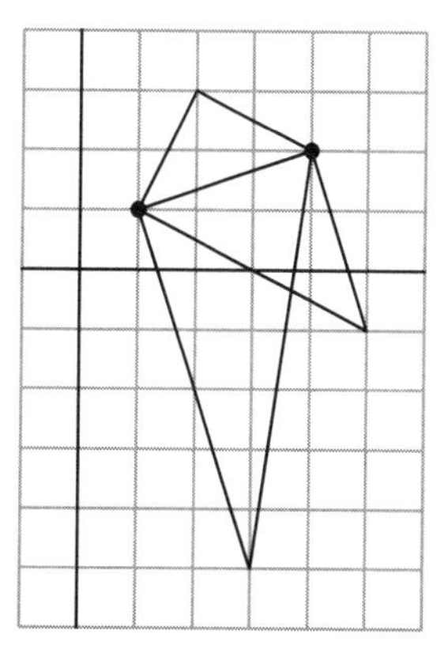

D. There are many possible pairs of vertices, such as (1, $^-$1) and (4, 0); (2, 4) and ($^-$1, 3); and (0, 2) and ($^-$3, 1).

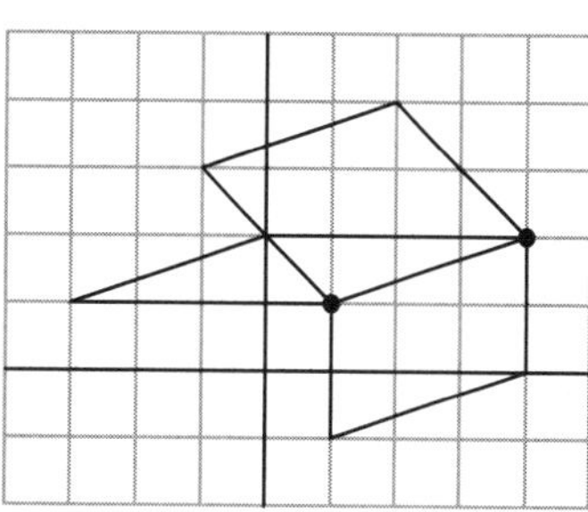

INVESTIGATION 2

Finding Areas and Lengths

In this investigation, students find areas of figures drawn on a dot grid and then explore the relationship between the area of a square and the length of its side.

In Problem 2.1, Finding Areas, students calculate areas of several figures drawn on a dot grid. In Problem 2.2, Looking for Squares, they search for all the squares that can by drawn on a 5-dot-by-5-dot grid. In the process, they will begin to see how the area of a square relates to its side length. In the follow-up, students are introduced to the concept of square root. In Problem 2.3, Finding Lengths, they find the lengths of all the line segments that can be drawn on a 5-dot-by-5-dot grid. Then, they develop a strategy for finding the distance between two points by analyzing the line segment between them: they draw a square using the segment as one side, find the area of the square, and then find the positive square root of that area.

Mathematical and Problem-Solving Goals

- *To find areas of polygons drawn on a dot grid using various strategies*
- *To find the length of a line segment drawn on a grid by thinking of it as the side of a square*
- *To begin to develop an understanding of the concept of square root*

Materials

Problem	For students	For the teacher
All	Graphing calculators	Transparencies: 2.1 to 2.3 (optional), overhead geoboard (optional)
2.1	Labsheet 2.1, dot paper, or geoboards (1 per pair)	Transparency of Labsheet 2.1 (optional), transparent dot paper (optional)
2.2	Labsheet 2.2 (1 per student), centimeter rulers (1 per student)	
2.3	Labsheet 2.3 (1 per student), centimeter rulers (1 per student), dot paper or geoboards (optional)	
ACE	Labsheet 2.ACE (optional; 1 per student), dot paper	

Finding Areas and Lengths

In the last investigation, you solved problems involving coordinate grids. You located points, calculated distances, and found coordinates of polygons that satisfied given conditions. In this investigation, you will find areas of figures and lengths of line segments drawn on a dot grid.

2.1 Finding Areas

In this problem, you will find areas of figures drawn on a dot grid. Consider the horizontal or vertical distance between two adjacent dots to be 1 unit. The area of a square with side lengths of 1 unit is 1 square unit.

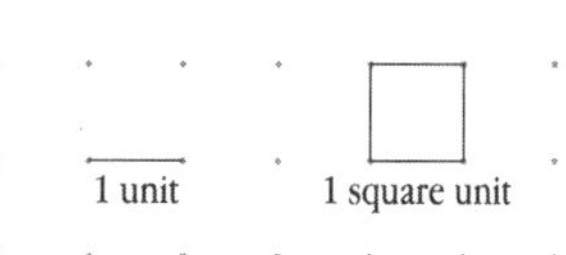

Problem 2.1

A. The figures below appear on Labsheet 2.1. Find the area of each figure below.

B. Describe the strategies you used to find the areas.

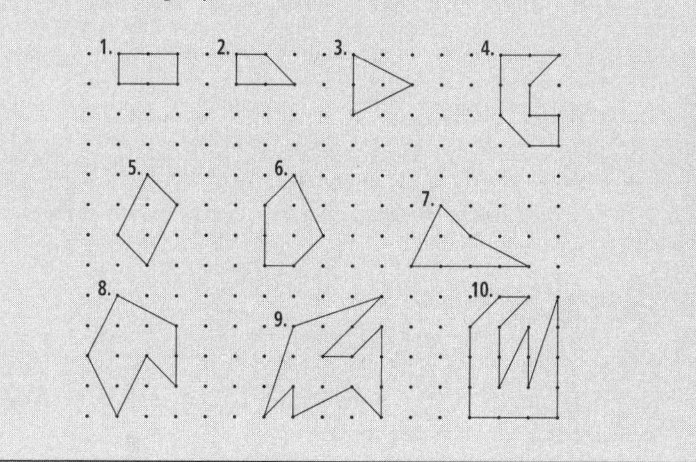

Answers to Problem 2.1

See page 26h.

2.1 Finding Areas

At a Glance

Grouping: *pairs*

Launch

- Introduce the idea of finding areas of figures on dot grids.
- Solicit ideas about how to find the area of one of the figures.
- Distribute geoboards, dot paper, or Labsheet 2.1.
- Have pairs work on the problem and follow-up (or, save the follow-up until you have summarized the problem).

Explore

- As pairs work, look for good strategies for finding areas of rectangles and triangles.
- Help students who need assistance applying the rule for the area of a triangle.

Summarize

- Help the class generalize strategies for finding areas of figures on dot grids.

Assignment Choices

ACE questions 1, 3, 8, 9, 13, 14, and unassigned choices from earlier problems

Looking for Squares

At a Glance

Grouping:
small groups

Launch

- Draw a unit square on a 5-dot-by-5-dot grid and label its area.
- Challenge students to find as many squares with different areas as they can.
- Have groups of two or three work on the problem and follow-up.

Explore

- Help students who have trouble finding tilted squares.
- Remind students who find the same square more than once to check the areas.

Summarize

- Have students share the squares they found until all eight are displayed.
- Talk about the side lengths of the squares and the connection to square roots.

■ **Problem 2.1 Follow-Up**

1. Find the area of each triangle below.
2. Describe the strategies you used to find the areas.

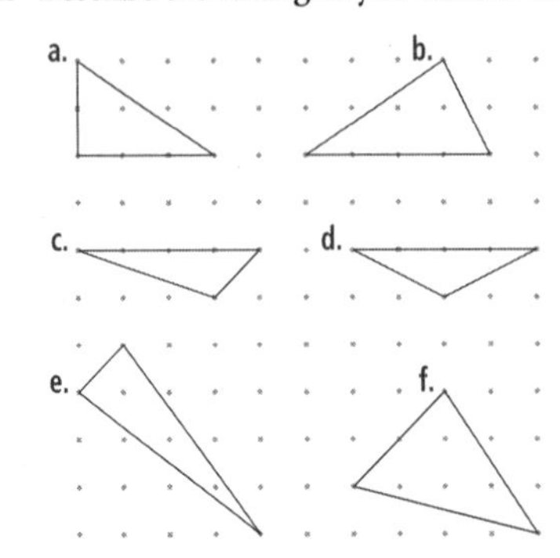

2.2 Looking for Squares

The smallest square you can draw by connecting the dots on a 5-dot-by-5-dot grid is a unit square, which has an area of 1 square unit. You can draw a 2-by-2 square by connecting the dots as shown. Since a 2-by-2 square contains 4 unit squares, it has an area of 4 square units.

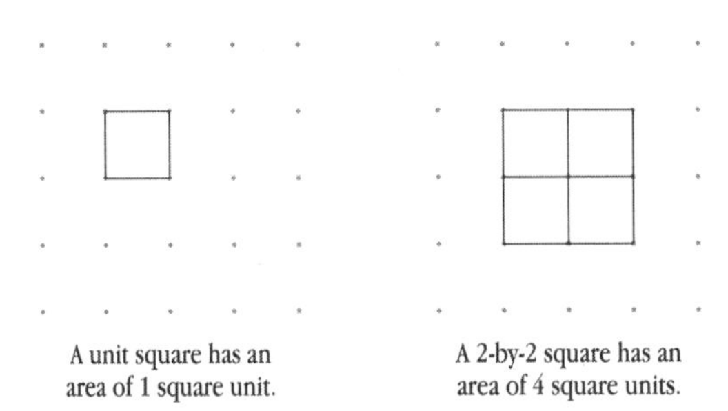

A unit square has an area of 1 square unit. A 2-by-2 square has an area of 4 square units.

In this problem, you will explore the other areas that are possible for squares drawn on a 5-dot-by-5-dot grid.

Problem 2.2

On the 5-dot-by-5-dot grids on Labsheet 2.2, draw squares of various sizes by connecting dots. Try to draw squares with as many different areas as possible. Label each square with its area.

Assignment Choices

ACE questions 4, 6, 7, 11, 15, and unassigned choices from earlier problems

Answers to Problem 2.1 Follow-Up

1. a. 3 square units b. 4 square units c. 2 square units
   d. 2 square units e. $3\frac{1}{2}$ square units f. 5 square units
2. See page 26i.

Answer to Problem 2.2

See page 26i.

Problem 2.2 Follow-Up

1. We will call squares with vertical and horizontal sides "upright" squares. Which of the squares you drew are upright squares? Identify each square by giving its area.
2. We will call squares with sides that are not vertical and horizontal "tilted" squares. Which of the squares you drew are tilted squares? Identify each square by giving its area.
3. For which kind of square—upright or tilted—is it easier to find the length of a side? Why?

The area of a square is the length of a side multiplied by itself. This can be expressed by the formula $A = s \times s$, or $A = s^2$. If you know the area of a square, you can work backward to find the length of a side. For example, suppose a square has an area of 4 square units. To find the length of a side, you need to figure out what positive number multiplied by itself equals 4. Since $2 \times 2 = 4$, the side length is 2 units. We call 2 a *square root* of 4.

In general, if $A = s^2$, then s is called a **square root** of A. Since $2 \times 2 = 4$ and $^-2 \times {^-2} = 4$, 2 and $^-2$ are both square roots of 4. Every positive number has two square roots. The symbol for the positive square root is $\sqrt{\ }$. We write $\sqrt{4} = 2$. The negative square root of 4 is $^-\sqrt{4} = {^-2}$.

4. **a.** What is the value of $\sqrt{1}$?
   b. What is the value of $\sqrt{9}$?
   c. What is the value of $\sqrt{16}$?
   d. What is the value of $\sqrt{25}$?
5. **a.** Is $\sqrt{2}$ greater than 1? Is $\sqrt{2}$ greater than 2? Explain your reasoning.
   b. The side length of a square with an area of 2 square units is $\sqrt{2}$ units. In Problem 2.2, you drew a square with an area of 2 square units. Use a centimeter ruler to find the side length of this square. You made your drawings on centimeter dot grids, so 1 centimeter = 1 unit.
   c. Use the square root button on your calculator to find $\sqrt{2}$. How does the answer compare to your answer to part b?

Answers to Problem 2.2 Follow-Up

1. Squares with area 1, 4, 9, and 16 square units are upright squares.
2. Squares with area 2, 5, 8, and 10 square units are tilted squares.
3. The side lengths of upright squares are easier to find because you can just count spaces between dots.
4. a. 1 b. 3 c. 4 d. 5
5. a. Since $\sqrt{1} = 1$ and $\sqrt{4} = 2$, $\sqrt{2}$ must be greater than 1 and less than 2.
   b. about 1.4 cm
   c. Possible answer: The calculator's answer, 1.414213562, is a bit more than I measured, but a ruler will not measure to this many decimal places.

Finding Lengths

At a Glance

Grouping:
small groups

Launch

- Draw a line segment on a 5-dot-by-5-dot grid and ask how its length might be found.
- Help the class make the connection between area and line segment length.
- Have groups of three or four explore the problem.

Explore

- Be sure everyone can complete a square on a line segment and find the segment's length.
- As groups finish, have them start on follow-up 1 and 2.

Summarize

- Have students share the lengths they found until all 14 are offered and describe their strategies for finding them.
- Talk about the follow-up and approximating square roots.

2.3 Finding Lengths

In Problem 2.2, you drew squares on 5-dot-by-5-dot grids. Some of the squares were upright and others were tilted. You learned that the positive square root of a number is the side length of a square with that number as its area. For example, $\sqrt{4} = 2$ because the square with an area of 4 square units has sides of length 2.

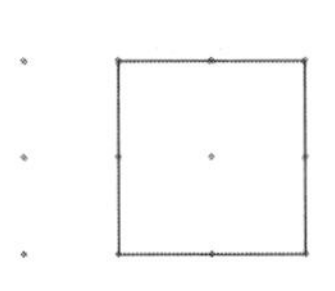

This square has an area of 4 square units. The length of a side is $\sqrt{4}$, or 2.

Sometimes a square root is not a whole number. For example, $\sqrt{2}$ is approximately 1.414. When a square root is not a whole number, we sometimes write it by using the $\sqrt{\ }$ symbol rather than finding a decimal approximation.

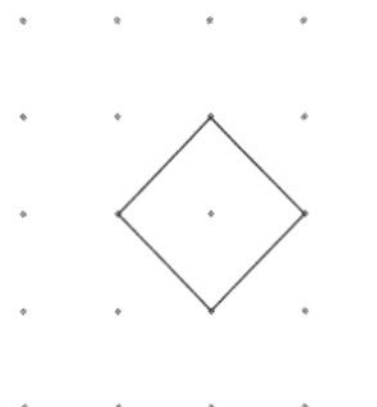

This square has an area of 2 square units. The length of a side is $\sqrt{2}$.

You can use a square to find the length of a line segment connecting dots on a grid. Just draw a square with the line segment as one side. The length of the segment is the square root of the square's area.

For example, to find the length of the line segment below, draw a square with the segment as one side. The square has an area of 5 square units, so the segment has a length of $\sqrt{5}$.

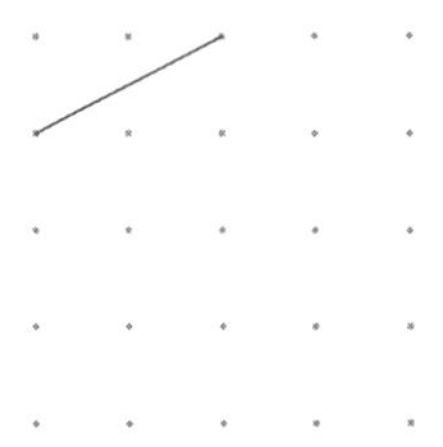

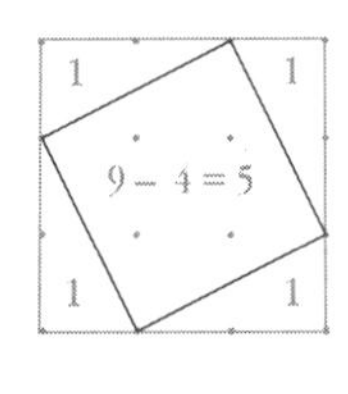

Assignment Choices

ACE questions 2, 5, 10, 12, and unassigned choices from earlier problems

Assessment

It is appropriate to use the check-up after this problem.

Answer to Problem 2.3

The possible lengths are 1, 2, 3, 4, $\sqrt{2}$, $\sqrt{8}$, $\sqrt{18}$, $\sqrt{32}$, $\sqrt{5}$, $\sqrt{20}$, $\sqrt{13}$, $\sqrt{25}$, $\sqrt{10}$, and $\sqrt{17}$. See the Summarize section for more information.

Answers to Problem 2.3 Follow-Up

1. a. Measurements will vary but should be reasonably close to those given in part b.

Problem 2.3

On the 5-dot-by-5-dot grids on Labsheet 2.3, draw line segments of various lengths by connecting dots. Try to draw segments with as many different lengths as possible. Use the method described on the previous page to find the length of each segment. To find some of the lengths, you will need to draw squares that extend beyond the 5-dot-by-5-dot grids. Label each segment with its length. Use the $\sqrt{\ }$ symbol to express lengths that are not whole numbers.

Problem 2.3 Follow-Up

1. a. Estimate each length that is not a whole number by measuring the segment with a centimeter ruler.
   b. Use your calculator to express each length that is not a whole number as a decimal to the thousandths place.
2. Express the length of the line segment below by using the $\sqrt{\ }$ symbol.

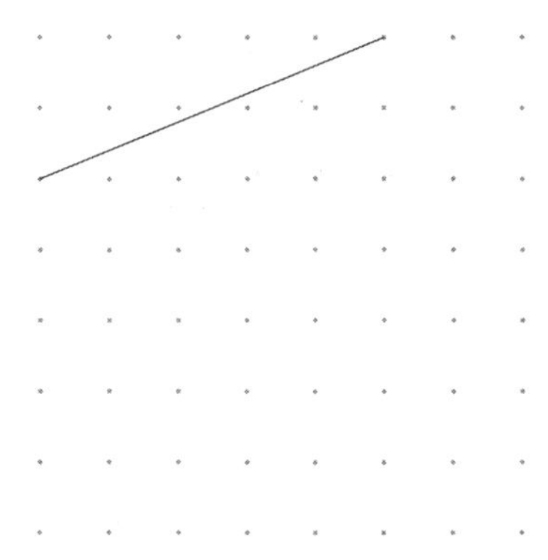

3. Between what two consecutive whole numbers is the length of the line segment in question 2?

b.

Length	Decimal approximation
$\sqrt{2}$	1.414
$\sqrt{8}$	2.828
$\sqrt{18}$	4.243
$\sqrt{32}$	5.657
$\sqrt{5}$	2.236
$\sqrt{20}$	4.472
$\sqrt{13}$	3.606
$\sqrt{10}$	3.162
$\sqrt{17}$	4.123

2, 3. See page 26i.

Answers

Applications

1. *figure a:* 5 square units, *figure b:* 2.5 square units, *figure c:* 1 square unit, *figure d:* 5.5 square units, *figure e:* 8.5 square units; Methods will vary. Students might subdivide the figure into smaller squares and triangles and add their areas. They might surround the figure with a rectangle and subtract the areas of the shapes lying outside of the figure from the rectangle's area. For example, a square of area 4 square units can be drawn around figure c and the area of the three 1-square-unit triangles subtracted, leaving an area of 1 square unit.

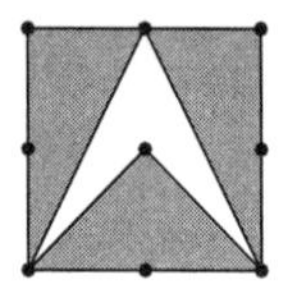

2. See below right.
3. See page 26j.
4. See page 26j.
5. See page 26j.

Applications • Connections • Extensions

As you work on these ACE questions, use your calculator whenever you need it.

Applications

In 1 and 2, refer to the figures below.

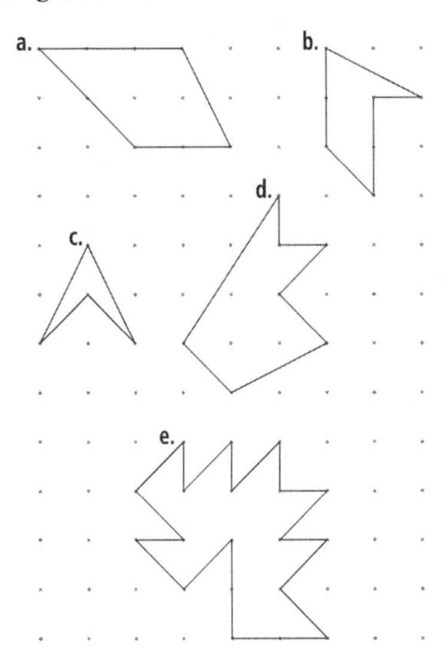

1. Find the area of each figure. Copy the figure onto dot paper if you need to. Describe the methods you use to find the areas.
2. For figures c and d, find the length of each side.
3. On dot paper, draw a hexagon with an area of 16 square units.
4. Find every possible area for a square drawn by connecting dots on a 3-dot-by-3-dot grid.
5. Find every possible length for a line segment drawn by connecting dots on a 3-dot-by-3-dot grid.

2.

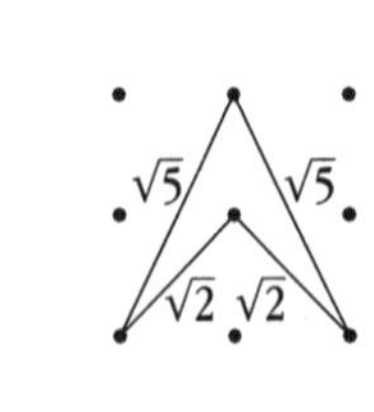

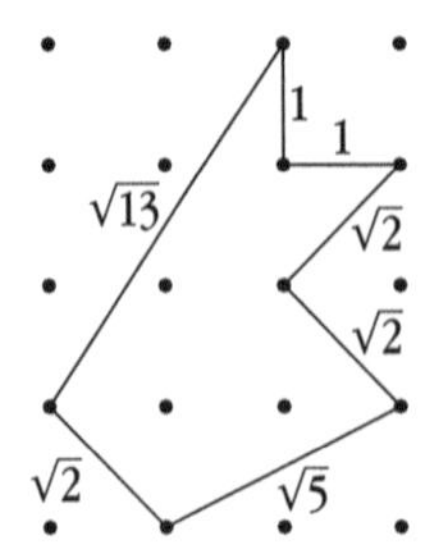

In 6 and 7, refer to the diagram below.

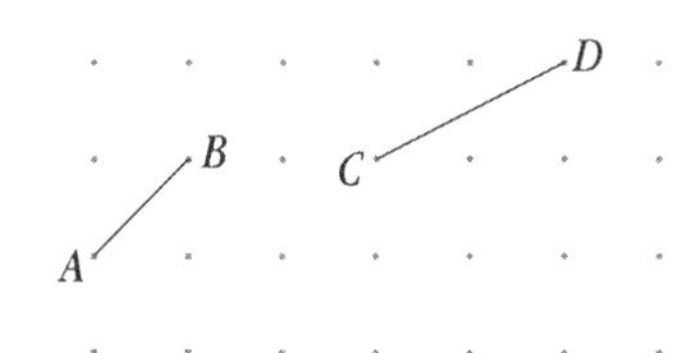

6. a. On dot paper, draw a square with segment *AB* as a side. What is the area of the square?

b. Use a calculator to estimate the length of segment *AB*.

7. a. On dot paper, draw a square with segment *CD* as a side. What is the area of the square?

b. Use a calculator to estimate the length of segment *CD*.

Connections

8. a. Find the areas of triangles *AST*, *BST*, *CST*, and *DST*. How do the areas compare? Why do you think this is true?

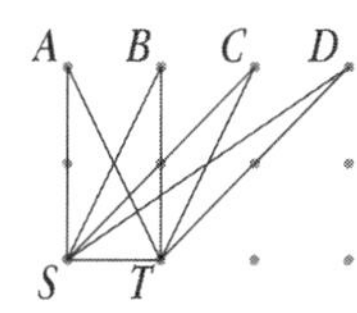

b. Find the areas of triangles *VMN*, *WMN*, *XMN*, *YMN*, and *ZMN*. How do the areas compare? Why do you think this is true?

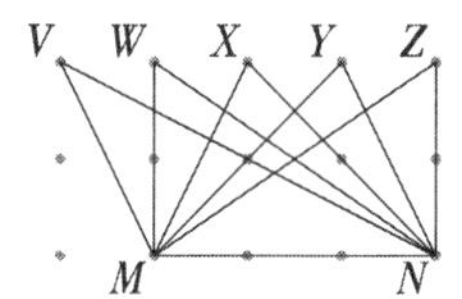

6a. 2 square units

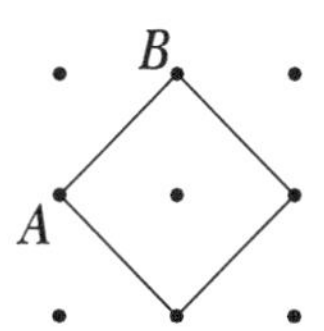

6b. about 1.414

7a. 5 square units

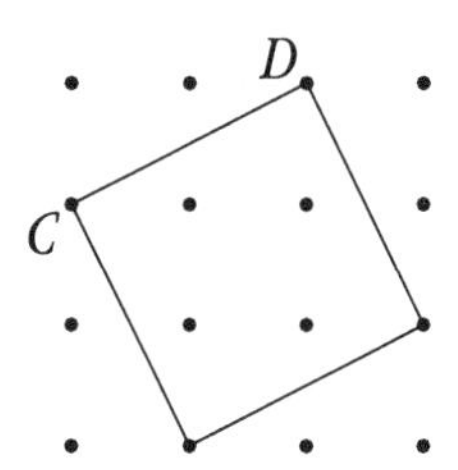

7b. about 2.236

Connections

8a. Each triangle has an area of 1 square unit. They all have base 1 and height 2, so $A = \frac{1}{2}bh = 1$ for each.

8b. Each triangle has an area of 3 square units. They all have base 3 and height 2, so $A = \frac{1}{2}bh = 3$ for each.

9a. Possible answer: By subdividing this square along its diagonals, you get four triangles, each with an area of $\frac{1}{2}$ square unit. Therefore, the square has an area of 2 square units.

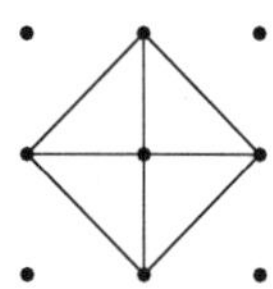

9b. The outer triangles each have an area of $\frac{1}{2}$ square unit. Together they have a combined area of 2 square units which, when subtracted from the upright square's area of 4 square units, leaves an area of 2 square units.

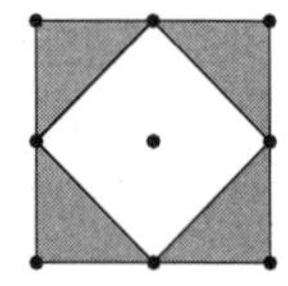

10a. The perimeter is the length of a side multiplied by 4.

10b. See page 26j.

11a. See below right.

9. a. On dot paper, draw a square with an area of 2 square units. Write an argument to convince a friend that the area of the square is 2 square units.

b. Now, draw your square from part a inside an upright square with an area of 4 square units. There will be four triangles around your square. Find the area of each of these triangles. How can you use the upright square to find the area of the smaller square?

10. a. What rule can you use to calculate the perimeter of a square if you know the length of a side?

b. Give the perimeter of each square you drew in Problem 2.2 to the nearest hundredth.

11. a. Which of the triangles below are right triangles? Use the corner of a piece of paper to help you find right angles. Copy the right triangles onto dot paper, and mark the right angles.

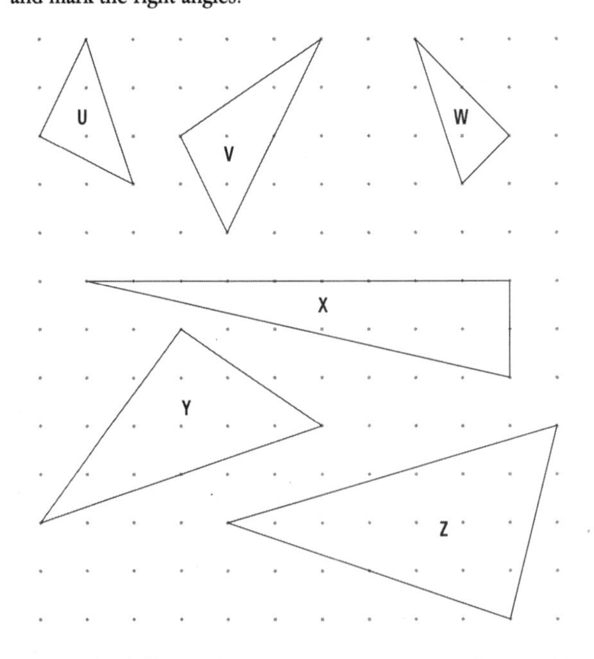

11a. Triangles U, W, and X are right triangles.

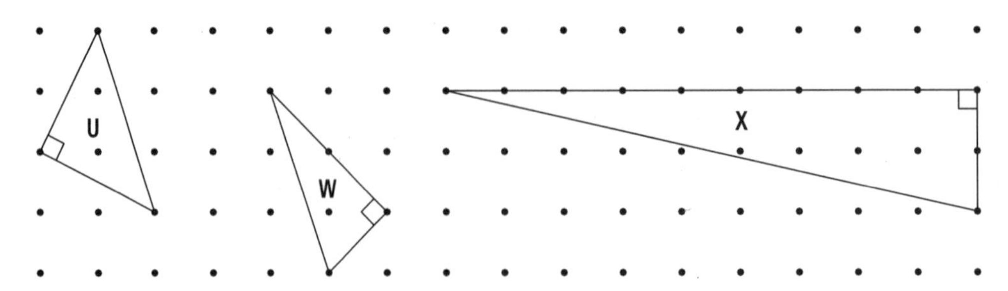

b. Find the area of each right triangle. Explain how you found the areas.

c. For each right triangle, find the sum of the measures of the two non–right angles. What do you notice? If you drew a different right triangle, what would be the sum of its non–right angles? How do you know?

Extensions

12. a. Three right triangles with a common side are drawn on the dot grid below. Find the length of the common side.

b. Do the three right triangles have the same area? Explain. Copy the figures onto dot paper if you need to.

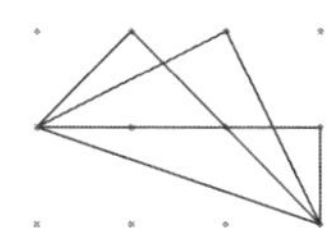

13. On dot paper, draw a parallelogram with an area of 6 square units.

14. On dot paper, draw a triangle with an area of 5 square units.

15. Give both square roots of each number.

a. 1 b. 4 c. 9

d. 16 e. 25 f. 2

11b. See below left.

11c. The measures of the two non–right angles have a sum of 90°. In any triangle, the sum of the three angle measures must be 180°, and subtracting 90° for the right angle leaves 90° for the other two angles to share.

Extensions

12. See page 26k.

13. See page 26k.

14. See page 26k.

15a. $^{-}1$, 1

15b. $^{-}2$, 2

15c. $^{-}3$, 3

15d. $^{-}4$, 4

15e. $^{-}5$, 5

15f. $^{-}\sqrt{2}$, $\sqrt{2}$ or $^{-}1.414$, 1.414

11b. *triangle U:* 2.5 square units, *triangle W:* 2 square units, *triangle X:* 9 square units; Possible explanation: I added a congruent triangle to make a rectangle. The area of each triangle is half the area of the corresponding rectangle.

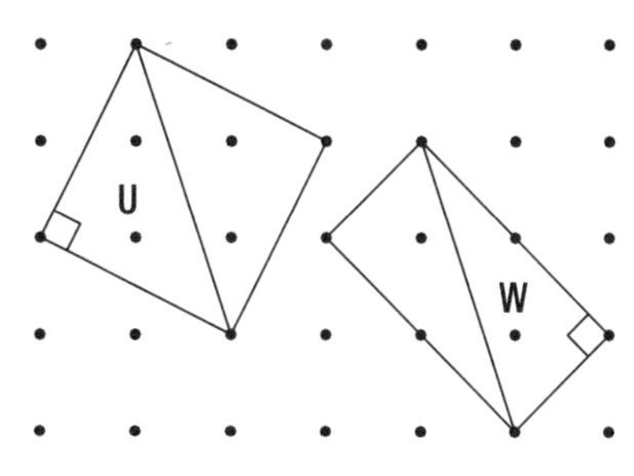

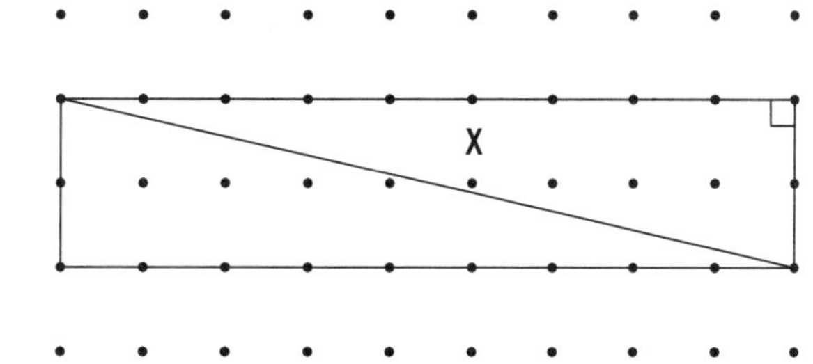

Possible Answers

1. We found the areas of figures drawn on dot paper in several ways. Sometimes we just counted the units of area. Sometimes we subdivided the figure into smaller shapes like right triangles and rectangles, counted the areas of the smaller shapes, and added them to get the large figure's area. And sometimes we enclosed the figure in a rectangle, found the area of the rectangle, and subtracted the areas of the figures that were not part of the enclosed figure.

2. To find the side length of an upright square drawn on dot paper, you can just count the units. To find the side length of a tilted square, you can find the area of the square and then take its positive square root (the positive number that when multiplied by itself equals the area).

3. See page 26l.

Mathematical Reflections

In this investigation, you explored areas of figures and lengths of segments drawn on dot paper. You learned that the side length of a square is the positive *square root* of the area. You discovered that, in many cases, a square root is not a whole number. These questions will help you summarize what you have learned:

1. Describe the strategies you used to find areas of figures drawn on dot paper. Give examples if it helps you to explain your thinking.
2. Describe how you would find the side length of a square drawn on dot paper without using a ruler. Consider both upright and tilted squares.
3. Describe how you would find the length of a line segment drawn on dot paper without using a ruler. Be sure to consider horizontal, vertical, and tilted segments.

Think about your answers to these questions, discuss your ideas with other students and your teacher, and then write a summary of your findings in your journal.

Tips for the Linguistically Diverse Classroom

Original Rebus The Original Rebus technique is described in detail in *Getting to Know Connected Mathematics.* Students make a copy of the text before it is discussed. During the discussion, they generate their own rebuses for words they do not understand; the words are made comprehensible through pictures, objects, or demonstrations. Example: Question 2—Key words and phrases for which students might make rebuses are *side length of a square* (a square with an arrow pointing to one side), *dot paper* (rectangle with dots inside), *without using a ruler* (ruler with an X over it), *upright* (upright square), *tilted* (tilted square).

TEACHING THE INVESTIGATION

2.1 • Finding Areas

In this problem, students find areas of a variety of figures drawn on a dot grid. They will begin to see that for some figures it is easy to find areas by subdividing and adding the areas of the component parts; other figures seem to need another approach.

Launch

Conduct the following short activity to introduce the idea of finding areas of figures drawn on a dot grid.

Draw a figure on a dot grid on the board, an overhead geoboard, or transparent dot paper. Choose a shape simple enough that students can easily find its area, either by subdividing it or enclosing it in a rectangle. For example:

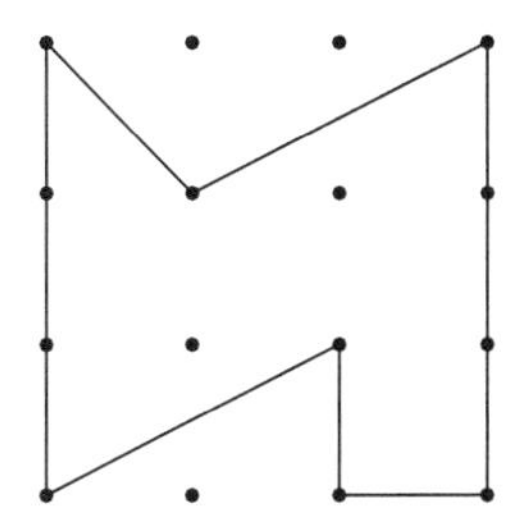

Ask:

How can you find the area of this figure?

Let students share their ideas. They will likely offer a version of the "divide the figure into smaller parts and add the separate areas" strategy. (It is not important that they completely work out their strategies at this point.) Here are two strategies that will work for most figures:

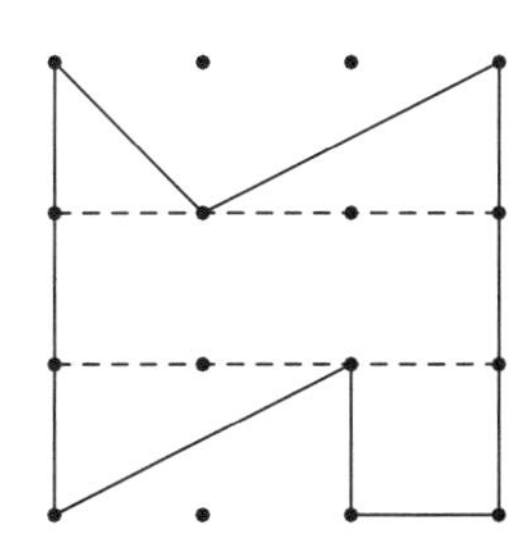

Subdivide the figure into simpler shapes, and add the areas of the simpler shapes.

area $= 3 + 3(1) + \frac{1}{2} = 6\frac{1}{2}$ square units

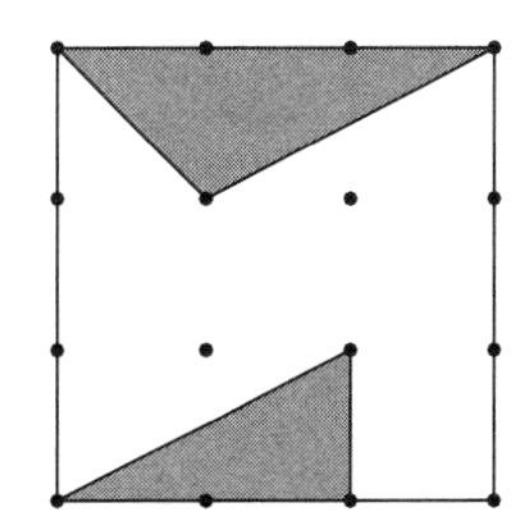

Enclose the figure in a rectangle, and subtract the areas of the figures outside the rectangle from the area of the rectangle.

area $= 9 - 1\frac{1}{2} - 1 = 6\frac{1}{2}$ square units

Have students explore the problem in pairs. All of the explorations can be done on dot paper or with geoboards. Labsheet 2.1 contains the figures for Problem 2.1 and its follow-up. Students may work on the labsheet, redraw the figures on dot paper, or construct them on geoboards.

For the Teacher: Using Geoboards to Explore Area

Many activities in this unit are classic geoboard problems of finding areas of figures. If you have access to geoboards, use them; students will enjoy exploring area with them. If your students have had experience with geoboards, this will go quickly. If not, spend some time helping them to become familiar with them. Demonstrate how to form shapes on a geoboard and how to use extra rubber bands to subdivide a figure or to surround it with a rectangle. You might have students pair up and create strange figures for each to find the area of. An overhead geoboard would also be helpful in this unit.

Explore

In their work, students will review how to find areas of rectangles and triangles. Look for students who are actively applying this knowledge; they can share their strategies in the summary. Students can count the number of units that cover the figure, or they can apply the rules for finding areas of rectangles and triangles. Some students may need help applying the rule for the area of a triangle, $A = \frac{1}{2}bh$. Help them to see that a triangle is half of a rectangle.

Assign the follow-up as pairs finish the problem, or assign it after the summary of the problem.

Summarize

As students share answers and strategies, help them to generalize their conclusions about finding area.

> We can find areas of some figures by subdividing them and adding the areas of the smaller figures. For which figures in this problem is using this method easy? *(Students will probably mention at least figures 1, 2, 3, and 4.)*

> We can find areas of some figures by enclosing them in a rectangle and subtracting the areas of the unwanted parts from the rectangle's area. For which figures in this problem is using this method easy? *(Students' ideas will vary. Figure 5, for example, can be enclosed in a 2-by-3 rectangle; the areas of four triangles—two with areas of $\frac{1}{2}$ square unit and two with areas of 1 square unit—are then subtracted from the rectangle's area, leaving 3 square units.)*

Some students may also use the strategy of rearranging parts of a figure to form a rectangle or a triangle with an easy-to-find area. For example, see the answer given for figure 3. Students will need to be able to apply these methods for their future work in this unit, so make sure everyone can find areas using at least one of these methods and can explain why the method works.

For the triangles in the follow-up, students may have applied the strategies reviewed above or used rules about the areas of triangles learned in earlier mathematics work.

2.2 • Looking for Squares

In this problem, students draw squares of various sizes on 5-dot-by-5-dot grids. In the process, they begin to see how the area of a square relates to the length of its sides. The concept of square root is introduced in the follow-up in the context of the relationship between the area of a square and the length of its sides.

Launch

Display Transparency 2.2, or draw a 5-dot-by-5-dot grid on the board. Draw a unit square on the grid, and label it with the numeral 1.

> I have drawn a square with area 1 on this 5-dot-by-5-dot grid. Can someone tell me how to draw a different square on a 5-dot-by-5-dot grid?

Explain that students are to search for all the different squares they can find that will fit on a 5-dot-by-5-dot grid. Distribute Labsheet 2.2 and have students work on the problem in groups of two or three.

Explore

Some students may find "upright" squares easily (such as a square with area 9 square units) but have difficulty finding "tilted" squares (such as a square with area 10 square units).

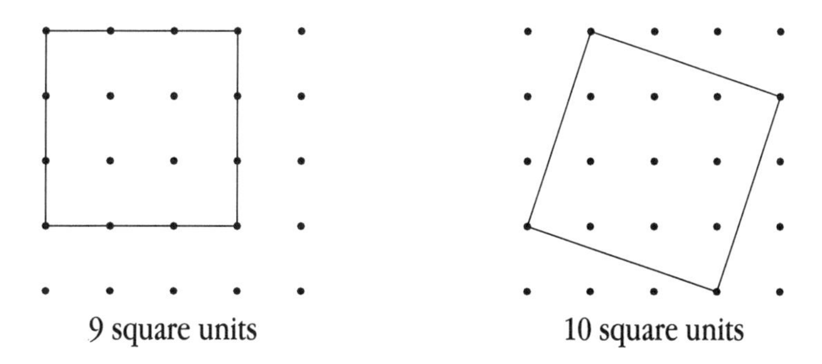

If students have difficulty identifying tilted squares, display one on the board or overhead. Start with a square of area 2.

If some students have found a square more than once, remind them to check the area of each square they draw to verify that they are different.

When students are reasonably sure they have found all the squares, distribute centimeter rulers and have them move on to the follow-up.

Summarize

Ask students to share the various squares they found as you draw them on Transparency 2.2. Continue until all eight different squares are displayed. (If students do not offer all eight, suggest the missing ones yourself.) Discuss the strategies that students used to find the squares.

Which squares were easy to find? Why? *(upright squares, because their sides align with the horizontal and vertical lines of dots in the grid)*

Which squares were not easy to find? Why? *(tilted squares, because their sides must meet at right angles but they do not align with horizontal and vertical lines of dots in the grid, so they do not have whole-unit measurements)*

Now discuss follow-up questions 1, 2, and 3. Question 3 asks for which kinds of squares it is easier to find the length of a side. Discuss the side lengths of some of the upright squares.

This square has an area of 4 square units. What is the length of a side? *(2)*

How do you know your answer is correct? *(You can easily count 2 units along any side, and $2 \times 2 = 4$, or $2^2 = 4$.)*

Next, talk about the side lengths of some of the tilted squares.

You said that this square has an area of 5 square units. How can you prove this? *(Subdivide the square into smaller regions and add their areas, or enclose the figure in a larger square and subtract the areas of the four triangles outside the figure from the area of the larger square.)*

What is the length of a side of this square?

Students may have measured with a centimeter ruler and found a side length of about 2.2. If no one has an answer, suggest that they measure the side with a ruler.

Is this correct? Does 2.2 squared equal 5? *(No, not exactly; 2.2 is too small.)* Try 2.3. Does 2.3 squared equal 5? *(No, 2.3 is too large.)*

We know that the area is 5 square units, but when we measure the side length we get only an approximate measure. Suppose we wanted to know the *exact* length of a side of this square. What would the *square* of that length have to be? *(It would have to be 5 because the area is 5 square units.)*

We have a word in mathematics for the length of a side of a square of area 5 square units. We call this length a *square root* of 5, and we write it $\sqrt{5}$. *(Write this on the board or overhead.)*

Continue the discussion about the concept of square root.

What number multiplied by itself is 4? *(2)* We can say this another way: The square root of 4 is the number that when squared, or multiplied by itself, is 4. We write this as $\sqrt{4}$.

Demonstrate this at the board or overhead.

We say that $\sqrt{4} = 2$, because $2 \times 2 = 4$.

Is there another number that you can multiply by itself to get 4? *(yes, $^-2$)*

In mathematics, the symbol $\sqrt{4}$ means a positive number. If we want the negative number, we write $^-\sqrt{4}$. Since we are working with area, we will be using the positive value of the square root.

So, what number multiplied by itself is 5? *($\sqrt{5}$)*

What is the length of a side of a square that has an area of 8 square units? *($\sqrt{8}$)*

Ask students for a few decimal approximations for $\sqrt{8}$. As a class, use a calculator to multiply each approximation by itself to check whether the result is 8. Then ask:

Do you think you can find a number that when multiplied by itself gives a result of 2?

Students' answers for parts b and c of question 5 will not equal exactly $\sqrt{2}$. If 1.4 is multiplied by itself, the answer is 1.96. If the number in the calculator's display, say 1.414213562, is multiplied by itself, it does not equal 2. Here, $1.414213562 \times 1.414213562 = 1.999999999$. Also, if you simply square the calculator's result, rather than re-entering the number, most calculators will tell you that 2 is the answer.

Help the class to summarize their findings. This is a good time to review with students how to find square roots on a calculator.

What are all the lengths of the sides of squares that you have found? *($1, \sqrt{2}, 2, \sqrt{5}, \sqrt{8}, 3, \sqrt{10}$, and 4)*

Question 5 leads into the topic of Problem 2.3, in which students are asked to find all the different lengths of line segments that can be drawn on a 5-dot-by-5-dot grid.

2.3 • Finding Lengths

In this problem, students develop a strategy for finding the distance between two points by examining the line segment between the points. To find the length of the line segment, they draw a square with the segment as one side; find the area of the square; and then find the square root of the area.

Launch

As a class, list all the side lengths that students have found so far in their work with 5-dot-by-5-dot grids: 1, $\sqrt{2}$, 2, $\sqrt{5}$, $\sqrt{8}$, 3, $\sqrt{10}$, and 4. Then ask:

> Can you draw a line segment on a 5-dot-by-5-dot grid with a length that is different from these?

Draw the segment the class suggests, or draw one of your own, on Transparency 2.3. For example:

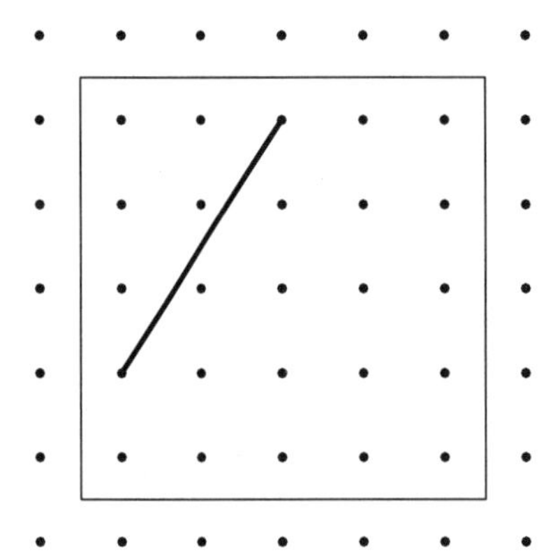

> How do you know that the length of this line segment is different from the others you have found?

Students might mention ways to informally measure the length of the segment, or they might suggest comparing the segment to others that are a bit shorter or longer.

> How might we find the actual length of this line segment?

Some students might suggest drawing a square using this line segment as a side and then calculating the segment's length from the square's area. If no one suggests this method, remind students of the connection they found between the area of a square and the length of a side.

The square that is drawn on each line segment in Problem 2.3 will extend beyond the 5-dot-by-5-dot grid.

For the line segment shown below, the area of the related square is 13 square units. Therefore, the length of a side, and of the segment, is $\sqrt{13}$.

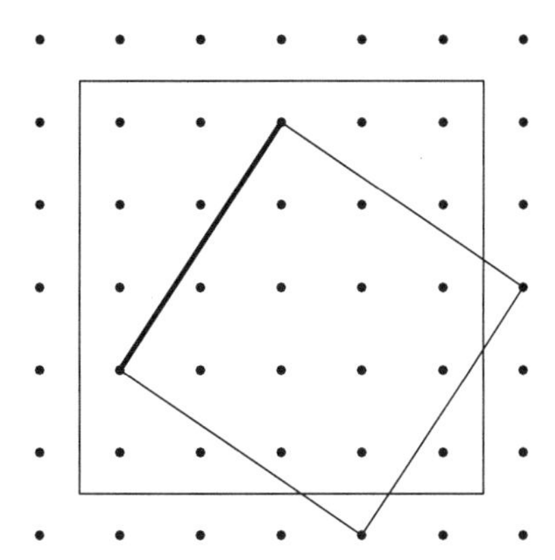

When students understand the process, have them explore the problem in groups of three or four. If geoboards are available, students can put two or more together to work on this problem. Or, they might draw the line segments on a larger dot grid.

Explore

Groups do not need to find all 14 possible lengths. However, be sure every student is able to complete a square on a line segment and find the length of the segment.

As students finish the problem, have them start on follow-up questions 1 and 2.

Summarize

Ask students to share the lengths they found, including the lengths of sides of the squares found in Problem 2.2. Draw the lengths on Transparency 2.3, or show them on an overhead geoboard. Continue until all 14 line segment lengths are displayed. (If students do not offer all 14, suggest the missing ones yourself.)

Arrange the lengths in an orderly way, perhaps as shown here:

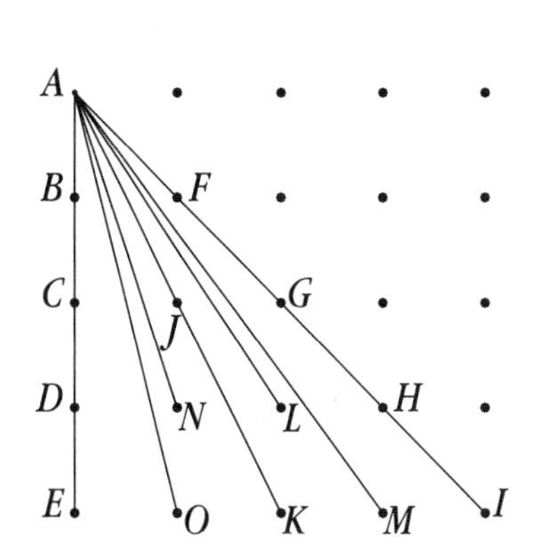

Line segment	Length
AB	1
AC	2
AD	3
AE	4
AF	$\sqrt{2}$
AG	$\sqrt{8}$ or $2\sqrt{2}$
AH	$\sqrt{18}$ or $3\sqrt{2}$
AI	$\sqrt{32}$ or $4\sqrt{2}$
AJ	$\sqrt{5}$
AK	$\sqrt{20}$ or $2\sqrt{5}$
AL	$\sqrt{13}$
AM	5 or $\sqrt{25}$
AN	$\sqrt{10}$
AO	$\sqrt{17}$

Discuss the strategies that students used to find the lengths. Some may have noticed relationships between line segments and didn't need to find a corresponding square. For example, segment *AG* above is twice the length of segment *AF*, or $2\sqrt{2}$. If others have found the length of *AG* by completing a square, they will have found $\sqrt{8}$. If your class is ready, talk about this equivalence, $\sqrt{8} = \sqrt{4 \times 2} = \sqrt{4} \times \sqrt{2} = 2\sqrt{2}$, or have students use a calculator to evaluate the various expressions to show their equivalence.

Before discussing the follow-up questions, ask students for approximations for some of the square roots they have found.

> Between what two whole numbers does $\sqrt{17}$ lie? *(4 and 5)* Which whole number is it closer to? *(It is closer to 4, because $4 \times 4 = 16$ and $5 \times 5 = 25$. A calculator tells us that it is about 4.123105626.)* Between what two whole numbers does $\sqrt{32}$ lie? *(5 and 6)*
>
> How many of the lengths we have listed would you have found on a 4-dot-by-4-dot grid? *(1, 2, and 3 as side lengths of upright squares; $\sqrt{2}$, $\sqrt{5}$, $\sqrt{8}$, $\sqrt{10}$, $\sqrt{13}$, and $\sqrt{18}$ as side lengths of tilted squares)*

Use follow-up question 3 as a final check of whether students have an appropriate strategy for finding the length of a line segment between two dots on a grid.

Additional Answers

Answers to Problem 2.1

A, B. Strategies will vary. A possible strategy is offered for each figure.

1.

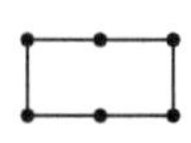

2 square units

2.

$2 - \frac{1}{2} = 1\frac{1}{2}$ square units

3.

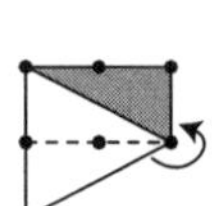

$1 \times 2 = 2$ square units

4.

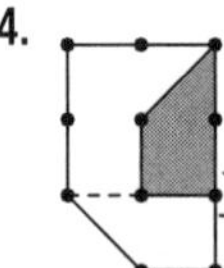

$2 \times 2 = 4$ square units

5.

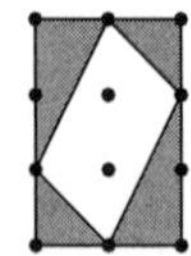

$6 - 2(1) - 2(\frac{1}{2}) = 3$ square units

6.

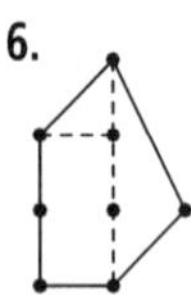

$2 + \frac{1}{2} + 1\frac{1}{2} = 4$ square units

7.

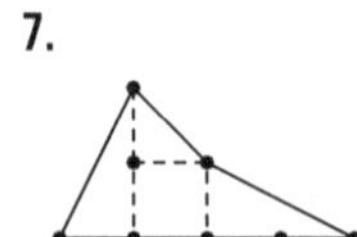

$3(1) + \frac{1}{2} = 3\frac{1}{2}$ square units

8.

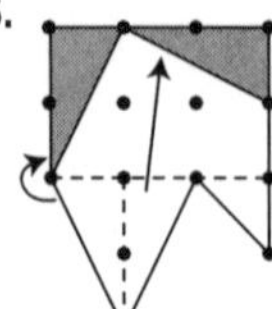

$6 + \frac{1}{2} = 6\frac{1}{2}$ square units

9.

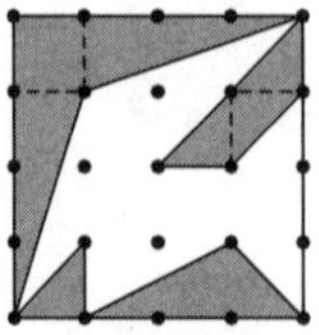

$16 - 1 - 3(1\frac{1}{2}) - 4(\frac{1}{2}) =$
$8\frac{1}{2}$ square units

10.

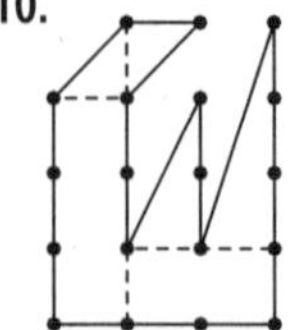

$3 + 2 + 2(\frac{1}{2}) + 1 + 1\frac{1}{2} =$
$8\frac{1}{2}$ square units

Answers to Problem 2.1 Follow-Up

2. The areas of triangles a–d can easily be found by thinking of each triangle as half a rectangle or by using the formula $\frac{1}{2}bh$. Students might enclose the triangles in rectangles and subtract the unwanted areas, particularly for triangles e and f:

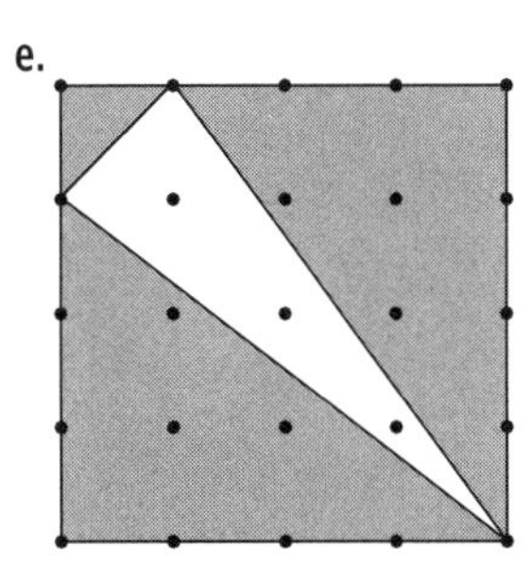

$16 - 6 - 6 - \frac{1}{2} = 3\frac{1}{2}$ square units

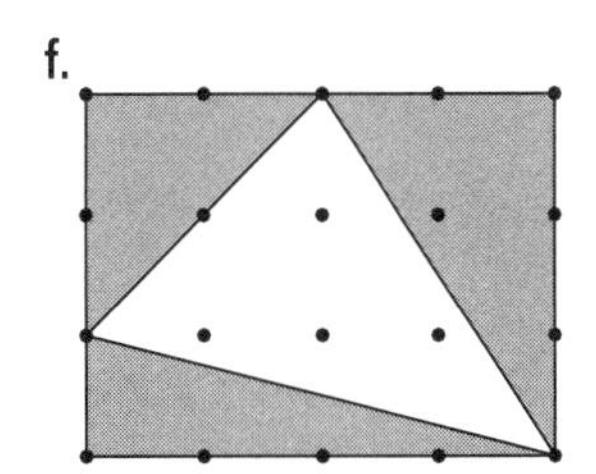

$12 - 2 - 3 - 2 = 5$ square units

Answer to Problem 2.2

Squares with eight different areas can be drawn:

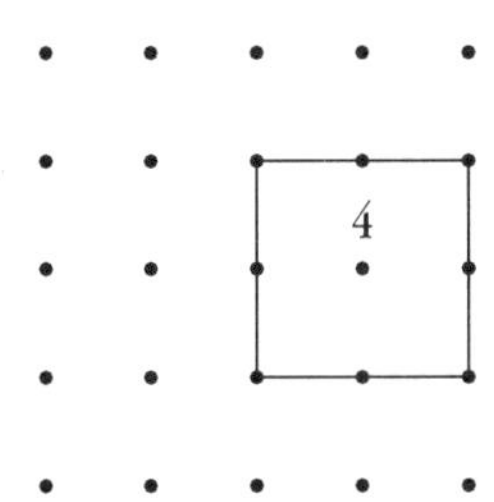

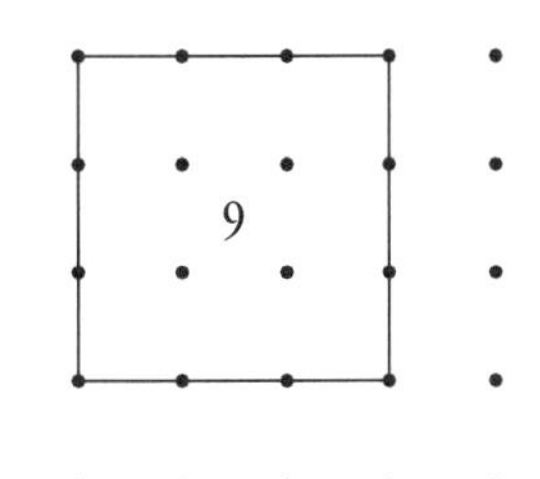

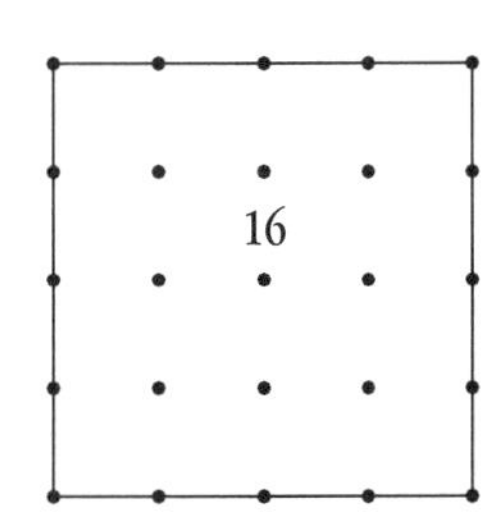

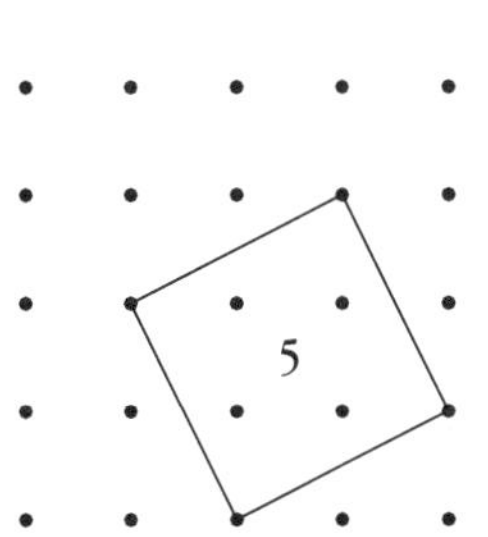

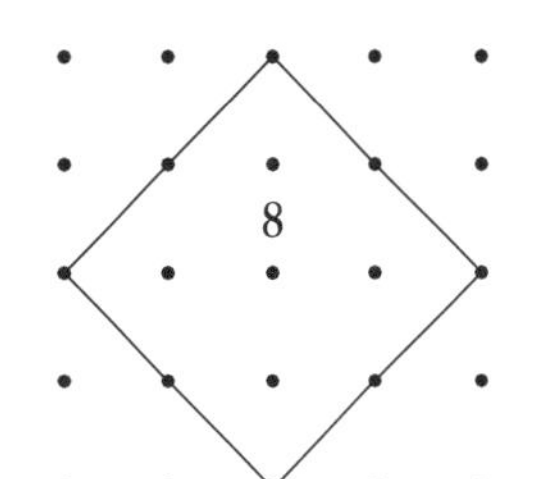

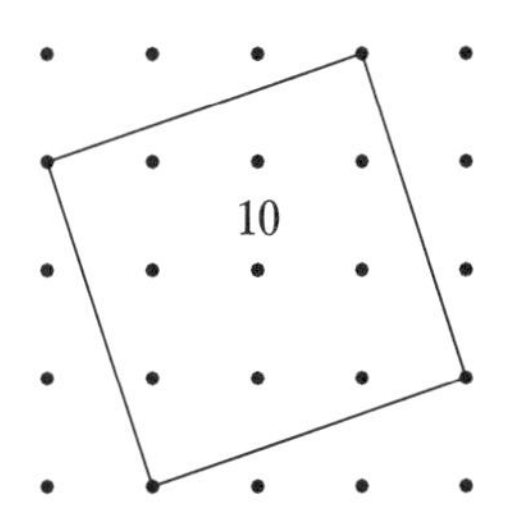

Answers to Problem 2.3 Follow-Up

2. The area of the square drawn with the line segment as a side is 49 – 20 = 29 square units, so the length of the segment is $\sqrt{29}$.

3. The length $\sqrt{29}$ is between 5 and 6, because $5 \times 5 = 25$ and $6 \times 6 = 36$ and 29 is between 25 and 36.

ACE Answers

Applications

3. Possible answers:

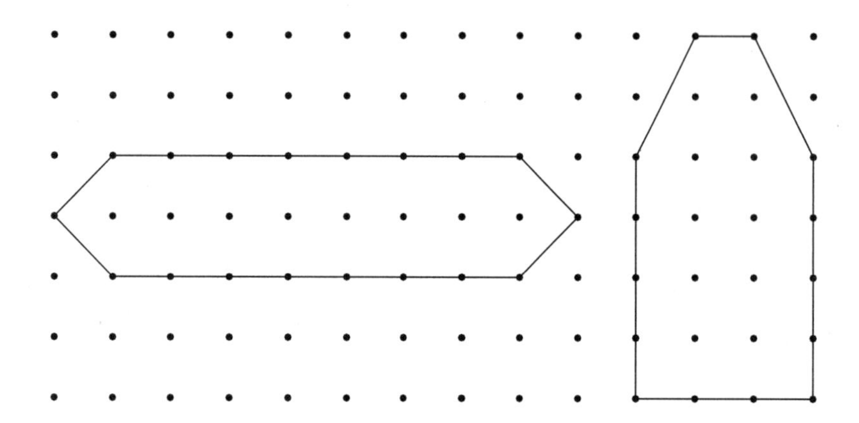

4. The possible areas are 1, 2, and 4 square units.

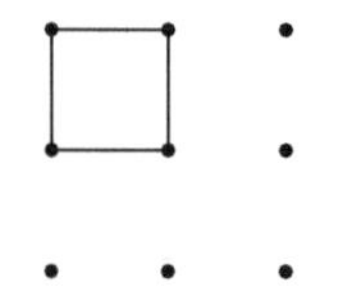

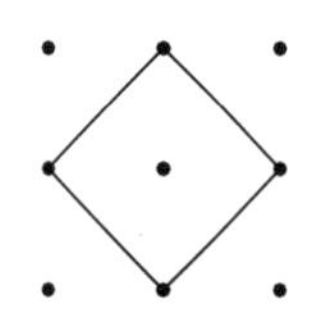

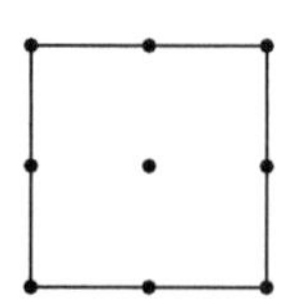

5. The possible lengths are 1, $\sqrt{2}$, 2, $\sqrt{5}$, and $\sqrt{8}$.

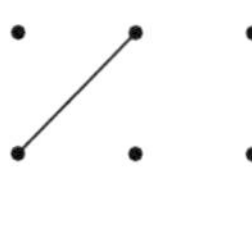

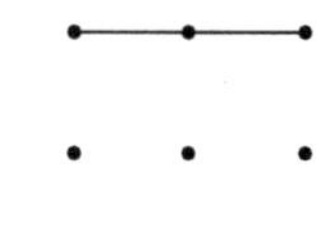

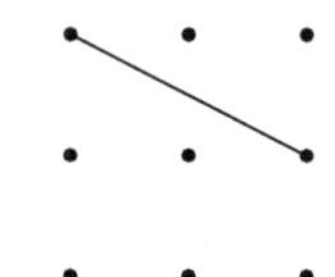

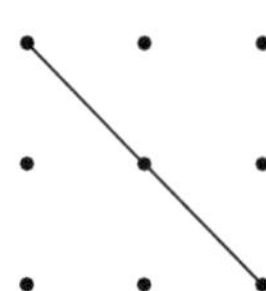

Connections

10b.

Area (square units)	Perimeter (units)
1	$4 \times 1 = 4.00$
2	$4 \times \sqrt{2} \approx 5.67$
4	$4 \times 2 = 8.00$
5	$4 \times \sqrt{5} \approx 8.94$
8	$4 \times \sqrt{8} \approx 11.31$
9	$4 \times 3 = 12.00$
10	$4 \times \sqrt{10} \approx 12.65$
16	$4 \times 4 = 16.00$

Teaching Tip: This is another opportunity to have students investigate between which whole numbers the square roots $\sqrt{2}$, $\sqrt{5}$, $\sqrt{8}$, and $\sqrt{10}$ lie. Even the observation that the lengths of the sides of the squares increase as the areas increase helps students to make sense of the size of these square roots.

Extensions

12a. A square drawn on the common side has an area of 10 square units, so the common side has length $\sqrt{10}$.

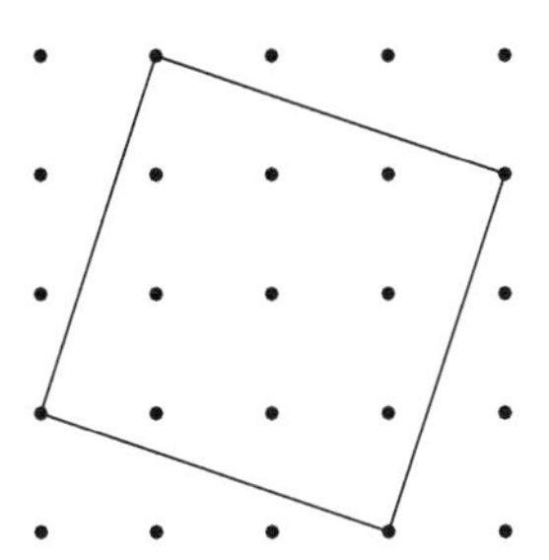

12b. The three right triangles do not have the same area. Possible explanations: The rectangles made from two triangles of the given area have different areas, so the triangles must have different areas.

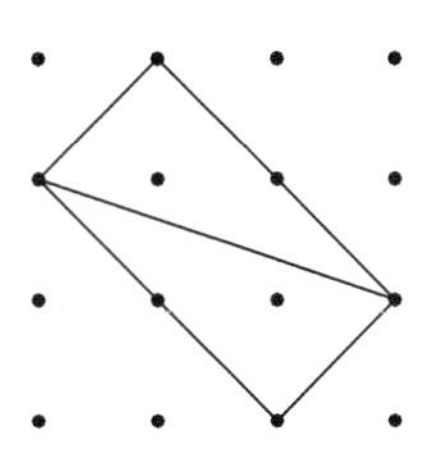

rectangle area = 4 square units

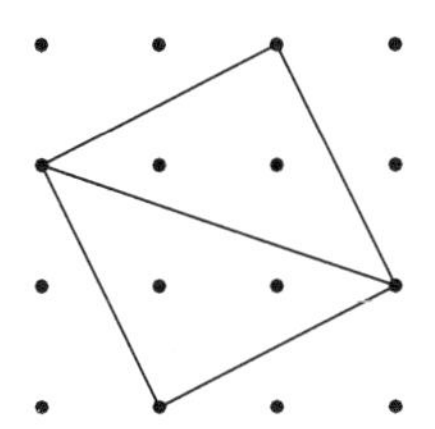

rectangle area = 5 square units

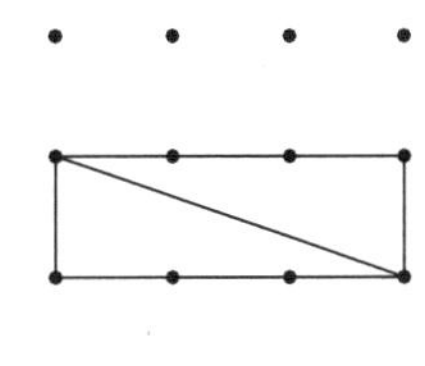

rectangle area = 3 square units

Or, as the three triangles have the same base but different heights, they must have different areas.

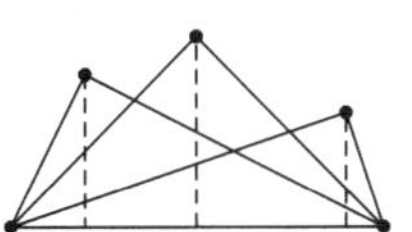

13. Possible answers:

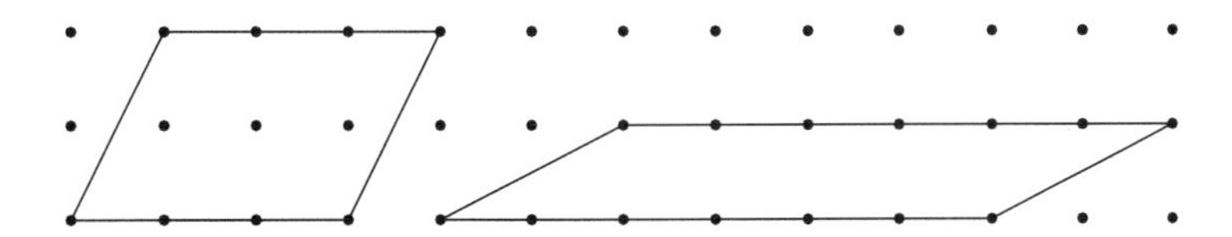

14. Possible answers:

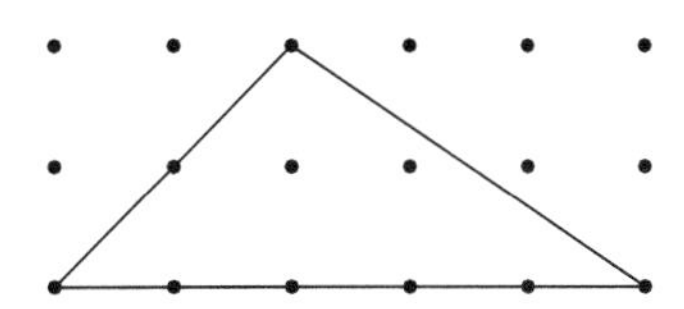

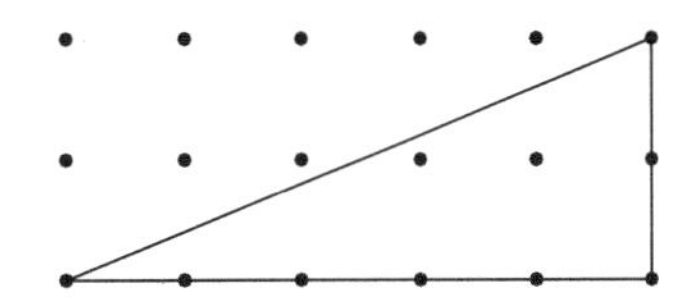

Mathematical Reflections

3. To find the length of a horizontal or vertical line segment drawn on dot paper, you can just count the units. To find the length of a diagonal line segment, you can draw a square with the segment as a side and then take the square root of the square's area, which gives the length of a side of the square. Or, you might be able to compare the segment to others for which you know the lengths. For example, the longer line segment below is twice the length of the shorter segment, which has a length of $\sqrt{2}$, so the longer segment is $2 \times \sqrt{2}$.

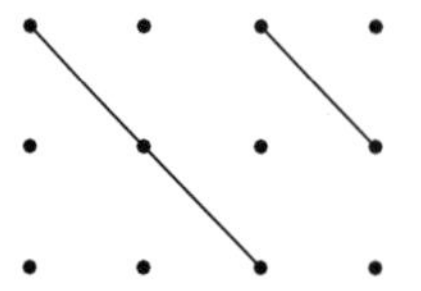

The Pythagorean Theorem

In this investigation, students discover the Pythagorean Theorem and explore its implications.

In Problem 3.1, Discovering the Pythagorean Theorem, students collect information about the areas of the squares on the sides of right triangles and conjecture that the sum of the areas of the two smaller squares equals the area of the largest square. In Problem 3.2, Puzzling Through a Proof, students investigate a puzzle that verifies that the sum of the areas of the squares on the legs of a right triangle is equal to the area of the square on the hypotenuse. In Problem 3.3, Finding Distances, they use the Pythagorean Theorem to find the distance between two dots on a grid. In Problem 3.4, Measuring the Egyptian Way, they determine whether a triangle is a right triangle by looking at the squares on the sides of the triangle: If a, b, and c are the lengths of the sides of a triangle and $a^2 + b^2 = c^2$, then the triangle is a right triangle.

Mathematical and Problem-Solving Goals

- ***To deduce the Pythagorean Theorem through exploration***
- ***To use the Pythagorean Theorem to find areas of squares drawn on a dot grid***
- ***To use the Pythagorean Theorem to find the distance between two points on a grid***
- ***To determine whether a triangle is a right triangle***
- ***To relate areas of squares to the lengths of the sides***

Materials

Problem	For students	For the teacher
All	Graphing calculators	Transparencies: 3.1 to 3.4B (optional)
3.1	Dot paper	
3.2	Labsheets 3.2A, 3.2B, and 3.2C (1 of one version per student), scissors	Puzzle pieces and frames cut from a transparency of Labsheet 3.2 (optional)
3.3	Labsheet 3.3 (1 per student), dot paper	Transparent grid (optional)
3.4	String or straws, rulers, scissors	Transparent grid (optional)

The Pythagorean Theorem

In the last investigation, you found areas of figures drawn on dot paper. To find these areas, you may have used a method that involved right triangles. Recall that a right triangle is a triangle with a right, or 90°, angle. The right angle of a right triangle is often marked with a square.

The longest side of a right triangle is the side opposite the right angle. We call this side the **hypotenuse** of the triangle. The other two sides are called the **legs.**

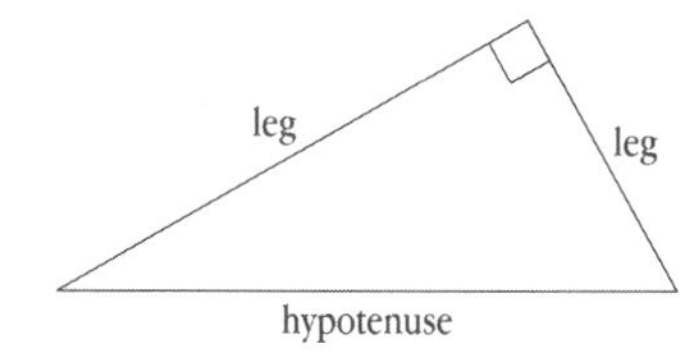

In this investigation, you will discover an interesting relationship among the side lengths of a right triangle.

3.1 Discovering the Pythagorean Theorem

Consider a right triangle with legs that each have a length of 1. Suppose you draw squares on the hypotenuse and legs of the triangle. How are the areas of these three squares related?

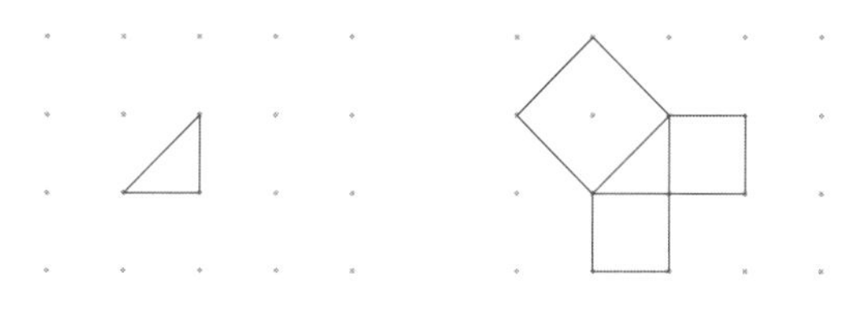

In this problem, you will look for a relationship among the areas of squares drawn on the sides of right triangles.

Discovering the Pythagorean Theorem

At a Glance

Grouping:
small groups

Launch

- Introduce the idea of drawing squares on the sides of a right triangle and comparing their areas.
- Describe the problem.
- Have groups of three or four work on the problem and follow-up.

Explore

- Ask that each student make a table.
- Encourage group members to share the work.

Summarize

- Talk about the pattern in the table.
- Discuss whether the relationship works for triangles that are not right triangles.
- Go over the follow-up.

Assignment Choices

ACE questions 3–7, 12, 13, 17, 18, and unassigned choices from earlier problems

Problem 3.1

A. Copy the table below. For each row, draw a right triangle with the given leg lengths on dot paper. Then, draw a square on each side of the triangle.

Length of leg 1	Length of leg 2	Area of square on leg 1	Area of square on leg 2	Area of square on hypotenuse
1	1	1	1	2
1	2			
2	2			
1	3			
2	3			
3	3			
3	4			

B. For each triangle, find the areas of the squares on the legs and on the hypotenuse. Record your results in the table.

C. Look for a pattern in the relationship among the areas of the three squares drawn for each triangle. Use the pattern you discover to make a conjecture about the relationship among the areas.

D. Draw a right triangle with side lengths that are different from those given in the table. Use your triangle to test your conjecture from part C.

Problem 3.1 Follow-Up

1. Add another column to your table. In the column, record the hypotenuse length of each triangle using the $\sqrt{\ }$ symbol.
2. Approximate each hypotenuse length to the hundredths place.

Answers to Problem 3.1

See page 40k.

Answers to Problem 3.1 Follow-Up

See page 40k.

3.2 Puzzling Through a Proof

The pattern you discovered in Problem 3.1 is a famous theorem named after the Greek mathematician Pythagoras. A *theorem* is a general mathematical statement that has been proven true. The Pythagorean Theorem is one of the most famous and important theorems in mathematics. Over 300 different proofs have been written for this theorem. One of these proofs is based on the geometric argument that you will explore in this problem.

Problem 3.2

Labsheet 3.2 contains two square puzzle frames and 11 puzzle pieces.

A. Cut out the puzzle pieces. Examine a triangular piece and the three square pieces. How do the side lengths of the squares compare to side lengths of the triangle?

B. Arrange the 11 puzzle pieces to fit exactly into the two puzzle frames. Use four triangles in each frame.

C. Carefully study the arrangements in the two frames. What conclusion can you draw about the relationship among the areas of the three square puzzle pieces?

D. What does the conclusion you reached in part C mean in terms of the side lengths of the triangles?

Compare your completed puzzles and your answers to parts C and D with those of another group.

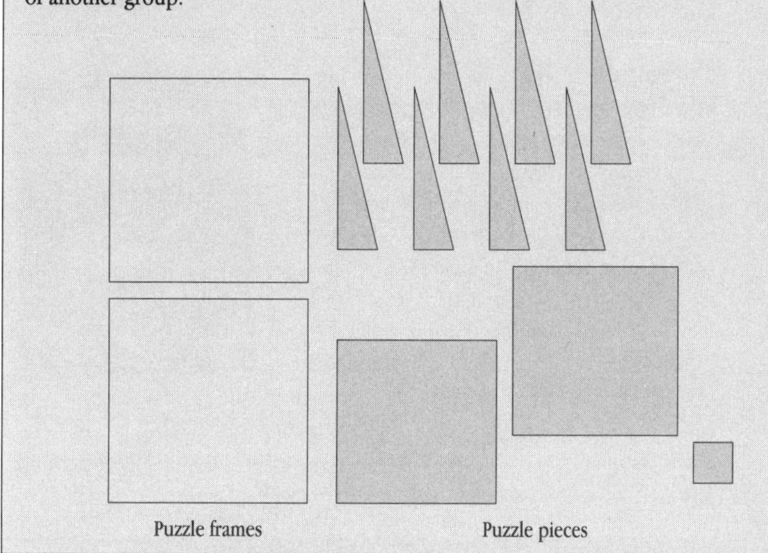

Puzzling Through a Proof

At a Glance

Grouping: *small groups*

Launch

- Display a set of puzzle pieces, and ask what relationships students see.
- Have groups of four work on the problem; save the follow-up until after the summary of the problem.

Explore

- Have each group work with one version of the puzzle.
- Encourage groups to find more than one way to arrange the pieces.

Summarize

- Ask students what relationships they notice among the puzzle pieces.
- Have a couple of groups show their completed puzzles, and help students to relate the puzzles to the theorem.
- Assign and then review the follow-up.

Answers to Problem 3.2

A. Each side length of the triangle is equal to the lengths of the sides of one of the three squares.

B. Possible arrangement:

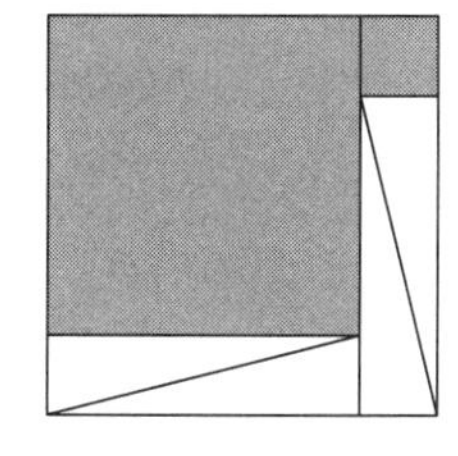

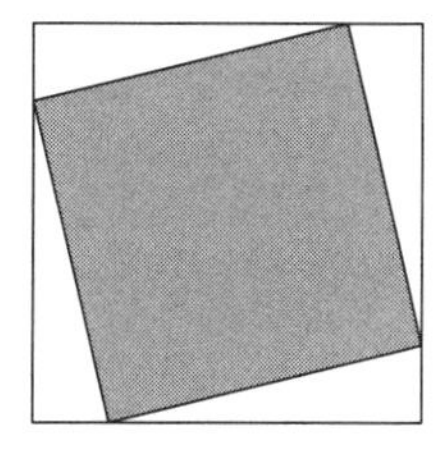

C. The sum of the areas of the two smaller squares is equal to the area of the largest square.

D. The sum of the squares of the lengths of the legs of a right triangle is equal to the square of the hypotenuse.

Assignment Choices

ACE questions 8–11, 20, and unassigned choices from earlier problems

Problem 3.2 Follow-Up

1. In Problems 3.1 and 3.2, you explored the Pythagorean Theorem. State this relationship as a general rule for any right triangle with legs of lengths a and b and a hypotenuse of length c.
2. A right triangle has legs of lengths 3 centimeters and 5 centimeters.
   - **a.** Use the Pythagorean Theorem to find the area of a square drawn on the hypotenuse of the triangle.
   - **b.** What is the length of the hypotenuse?
3. A right triangle has legs of lengths 5 inches and 12 inches.
   - **a.** Find the area of a square drawn on the hypotenuse of the triangle.
   - **b.** What is the length of the hypotenuse?
4. Use the Pythagorean Theorem to find the length of the hypotenuse of this triangle.

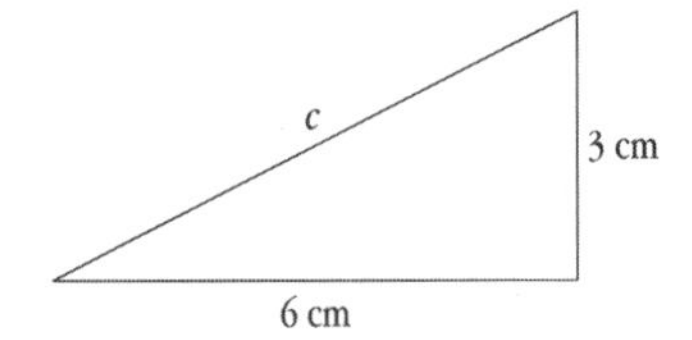

5. The hypotenuse of a right triangle is 15 centimeters long, and one leg is 9 centimeters long. How long is the other leg?

Did you know?

Pythagoras, a Greek mathematician who lived in the sixth century B.C., had a devoted group of followers known as the Pythagoreans. The Pythagoreans had many rituals, and they approached mathematics with an almost religious intensity. Their power and influence became so strong that some people feared that they threatened the local political structure, so they were forced to disband. However, many Pythagoreans continued to meet in secret and to teach Pythagoras's ideas.

Since they held Pythagoras in such high regard, the Pythagoreans gave him credit for all of their discoveries. Much of what we now attribute to Pythagoras, including the Pythagorean Theorem, may actually be the work of his followers.

Answers to Problem 3.2 Follow-Up

1. If a and b are the lengths of the legs of a right triangle and c is the length of the hypotenuse, then $a^2 + b^2 = c^2$.
2. a. $3^2 + 5^2 = 34$ cm^2
   b. $\sqrt{34} \approx 5.83$ cm
3. a. $5^2 + 12^2 = 169$ in^2
   b. $\sqrt{169} = 13$ in
4. Since $c^2 = 3^2 + 6^2 = 45$ cm^2, $c = \sqrt{45} \approx 6.7$ cm.
5. Since $15^2 - 9^2 = 144$ cm^2, the other leg is $\sqrt{144} = 12$ cm long.

3.3 Finding Distances

In Problem 2.3, you used squares to help you find the lengths of line segments connecting dots on a grid. The Pythagorean Theorem can also help you find these lengths.

Problem 3.3

A. 1. On the grid on Labsheet 3.3, draw a line segment connecting points *A* and *B*. Draw a right triangle with segment *AB* as its hypotenuse.

2. Find the lengths of the legs of the triangle.

3. Use the Pythagorean Theorem to find the length of the hypotenuse of the triangle.

B. Use the method described in part A to find the distance between points *C* and *D*.

C. Use the method described in part A to find the distance between points *E* and *F*.

B D
A
C
F
E

Problem 3.3 Follow-Up

On a sheet of dot paper, find two points that are $\sqrt{13}$ units apart. Label the points *X* and *Y*. Explain how you know that the distance between the points is $\sqrt{13}$.

3.3 Finding Distances

At a Glance

Grouping: ***pairs***

Launch

- Draw line segment *AB* as shown in the student edition, and ask how students could find its length.
- Do part A as a class.
- Have pairs work on the problem and follow-up.

Explore

- If students have trouble with the follow-up, ask questions to guide their thinking.

Summarize

- Ask students to explain their solutions to the problem.
- Go over the follow-up question carefully.
- Help the class visualize how the Pythagorean relationship can be used to find the distance between two dots on a grid.

Answers to Problem 3.3

See page 40l.

Answer to Problem 3.3 Follow-Up

Since $(\sqrt{13})^2 = 2^2 + 3^2$, the hypotenuse of a right triangle with legs of lengths 2 and 3 will have a length of $\sqrt{13}$.

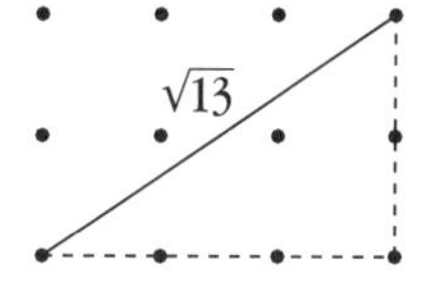

Assignment Choices

ACE questions 1, 2, 19, 21–26, and unassigned choices from earlier problems

Measuring the Egyptian Way

At a Glance

Grouping: *pairs*

Launch

- Talk about the two questions that introduce the problem.
- Have pairs work on the problem and follow-up.

Explore

- Help students who have trouble finding a triangle for parts A and B.
- Challenge some students to explore larger triangles.

Summarize

- Have students demonstrate the triangles they made.
- As a class, investigate multiples of side lengths of right triangles.
- Go over the follow-up.

3.4 Measuring the Egyptian Way

In this investigation, you have discovered the Pythagorean Theorem, which describes a special relationship among the side lengths of a right triangle. The theorem states that if a right triangle has legs of lengths a and b and a hypotenuse of length c, then $a^2 + b^2 = c^2$. In this problem, you will explore these questions:

- Is any triangle whose side lengths satisfy the relationship $a^2 + b^2 = c^2$ a right triangle?
- If the side lengths of a triangle do not satisfy the relationship $a^2 + b^2 = c^2$, does this mean the triangle is not a right triangle?

Problem 3.4

A. 1. Do the whole-number lengths 3, 4, and 5 satisfy the relationship $a^2 + b^2 = c^2$?

2. Form a triangle using string or straws cut to these lengths.

3. Is the triangle you formed a right triangle?

4. Repeat parts 1–3 with the lengths 5, 12, and 13.

5. Make a conjecture about triangles whose side lengths satisfy the relationship $a^2 + b^2 = c^2$.

B. 1. Form a triangle with side lengths a, b, and c that do not satisfy the relationship $a^2 + b^2 = c^2$.

2. Is the triangle a right triangle?

3. Repeat parts 1 and 2 with a different triangle.

4. Make a conjecture about triangles whose side lengths do not satisfy the relationship $a^2 + b^2 = c^2$.

Problem 3.4 Follow-Up

1. Determine whether the triangle with the given side lengths is a right triangle.

a. 12, 16, 20

b. 8, 15, 17

c. 12, 9, 16

Assignment Choices

ACE questions 14–16, 27, and unassigned choices from earlier problems

Answers to Problem 3.4

See page 40l.

Answers to Problem 3.4 Follow-Up

1. a. This is a right triangle, since $12^2 + 16^2 = 20^2$.

   b. This is a right triangle, since $8^2 + 15^2 = 17^2$.

   c. This is not a right triangle, since $12^2 + 9^2 \neq 16^2$.

In ancient Egypt, the Nile River overflowed every year, flooding the surrounding lands and destroying property boundaries. As a result, the Egyptians had to remeasure their land every year. Since many plots of lands were rectangular, the Egyptians needed a reliable way to mark right angles. They devised a clever method involving a rope with equally spaced knots.

2. Mark off 12 segments of the same length on a piece of rope or string. Tape the ends of the string together to form a closed loop. Try to form a right triangle with side lengths that are whole numbers of segments. You may need to have a classmate hold the string in place while you check that your triangle is a right triangle. What are the side lengths of the triangle you formed? Do these side lengths satisfy the relationship $a^2 + b^2 = c^2$?

3. How do you think the Egyptians used the knotted rope?

2. The side lengths of the triangle are 3, 4, and 5. They do satisfy the relationship $a^2 + b^2 = c^2$, since $3^2 + 4^2 = 5^2$.

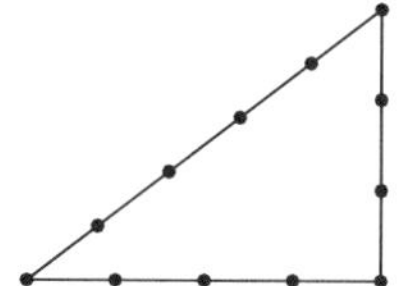

3. Possible answer: The Egyptians could have used the knotted rope to re-mark the property boundaries, perhaps setting one leg of the right triangle along the river bank and putting stakes in the ground along the other leg, which extended from the river bank at a 90° angle.

Answers

Applications

1. See below right.
2. See below right.
3. $h^2 = 4^2 + 3^2 = 25$, so $h = \sqrt{25} = 5$
4. $k^2 = 8^2 + 3^2 = 73$, so $k = \sqrt{73}$
5. $x^2 = 5^2 + 4^2 = 41$, so $x = \sqrt{41}$; $y = 20^2 + 4^2 = 416$, so $y = \sqrt{416}$
6. $d^2 = 30^2 - 10^2 = 800$, so $d = \sqrt{800}$
7. $j^2 = 12^2 - 4^2 = 128$, so $j = \sqrt{128}$

Applications • Connections • Extensions

As you work on these ACE questions, use your calculator whenever you need it.

Applications

1. On dot paper, find two points that are $\sqrt{17}$ units apart. Label the points W and X. Explain how you know that the distance between the points is $\sqrt{17}$.
2. On dot paper, find two points that are $\sqrt{20}$ units apart. Label the points Y and Z. Explain how you know that the distance between the points is $\sqrt{20}$.

In 3–7, find the missing length or lengths.

3.

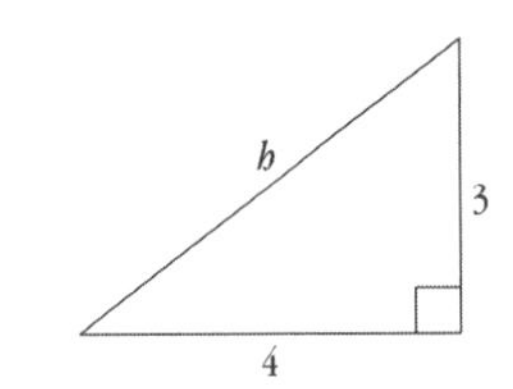

4.

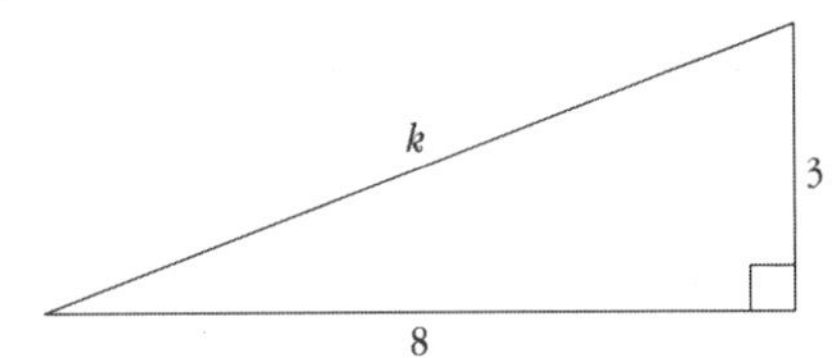

5.

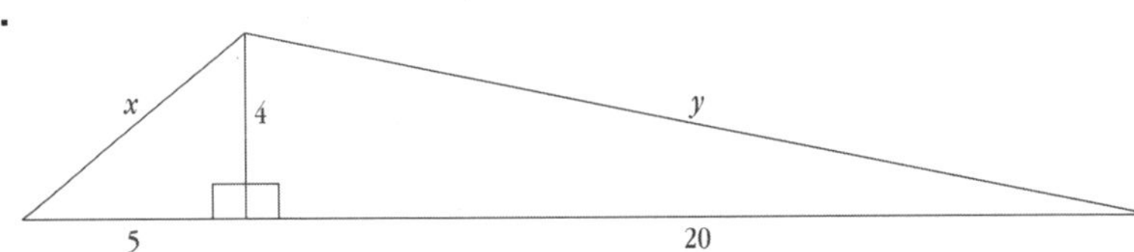

6.

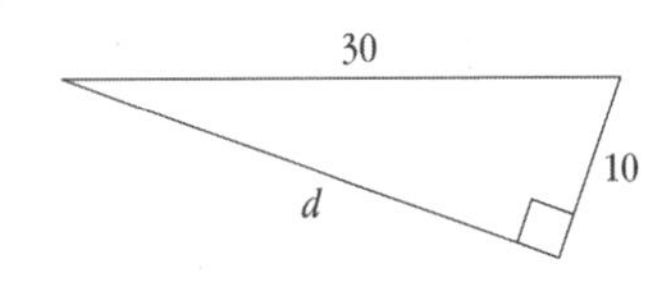

7.

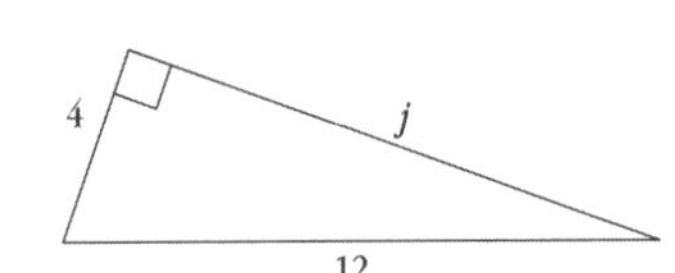

1. A right triangle can be formed with segment WX as the hypotenuse. Since $4^2 + 1^2 = 17$, the length of the hypotenuse and thus of segment WX is $\sqrt{17}$.

2. A right triangle can be formed with segment YZ as the hypotenuse. Since $4^2 + 2^2 = 20$, the length of the hypotenuse and thus of segment YZ is $\sqrt{20}$.

In 8–11, find the helicopter distance between the two locations in blocks without using a ruler, and explain how you found your answer.

8. the greenhouse and the stadium

9. the police station and the animal shelter

10. the greenhouse and the hospital

11. City Hall and the gas station

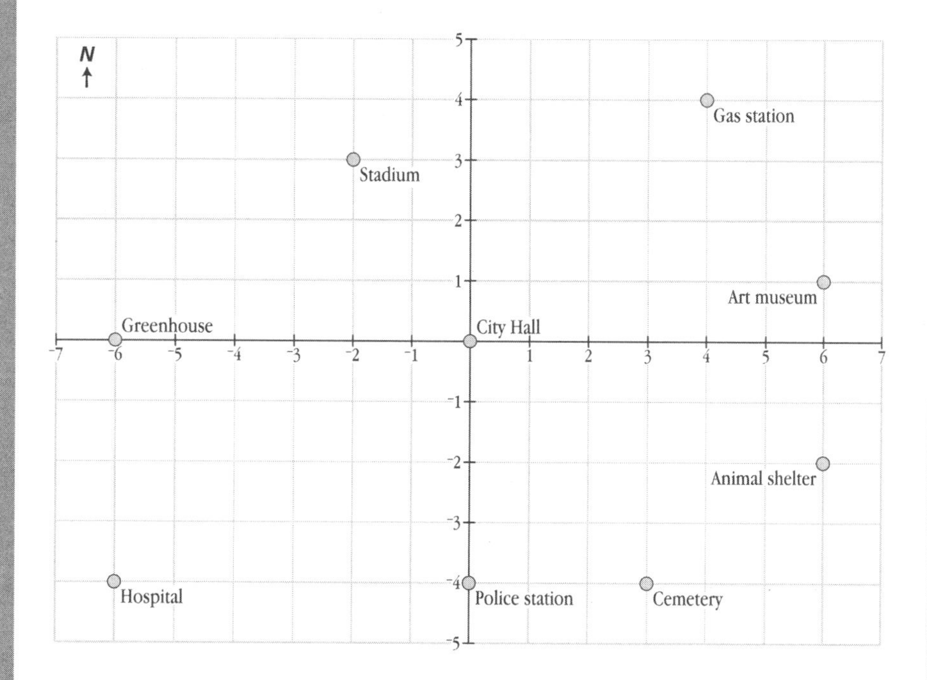

Note: In 8, 9, and 11, the distance between the two landmarks is found by thinking of that distance as the length of the hypotenuse of a right triangle.

8. Since $4^2 + 3^2 = 25$, the distance is $\sqrt{25} = 5$ blocks.

9. Since $6^2 + 2^2 = 40$, the distance is $\sqrt{40} \approx 6.3$ blocks.

10. This distance is 4 blocks.

11. Since $4^2 + 4^2 = 32$, the distance is $\sqrt{32} \approx 5.7$ blocks.

12. The squares on the legs each have an area of 2 square units, giving a sum of 4 square units—which is the same as the area of the square on the hypotenuse.

13. See page 40m.

14. This is a right triangle because $10^2 + 24^2 = 26^2$.

15. This is a right triangle because $10^2 + 10^2 = (\sqrt{200})^2$.

16. This is not a right triangle because $9^2 + 16^2 \neq 25^2$.

12. In this diagram, the squares on the legs of the right triangle are tilted, and the square on the hypotenuse is upright. Show that this triangle satisfies the Pythagorean Theorem. Copy the diagram onto dot paper if you need to.

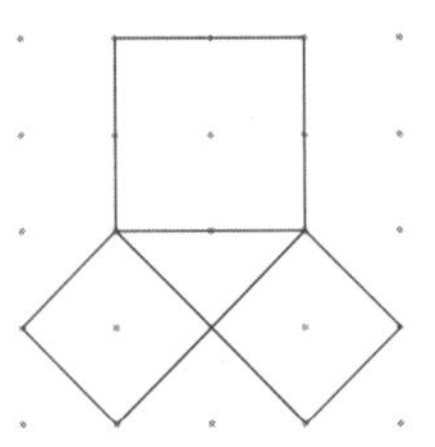

13. If you drew squares on the sides of this right triangle, they would all be tilted. Show that this triangle satisfies the Pythagorean Theorem. Copy the diagram onto dot paper if you need to.

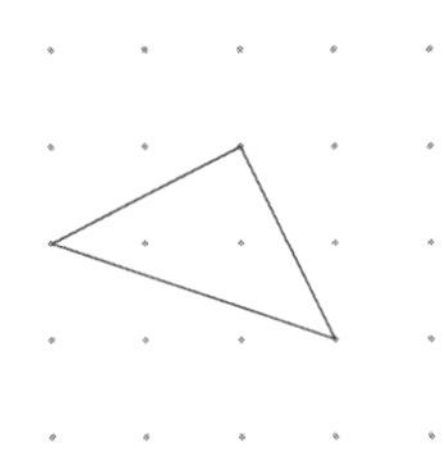

In 14–16, tell whether the triangle with the given side lengths is a right triangle.

14. 10, 24, 26

15. 10, 10, $\sqrt{200}$

16. 9, 16, 25

Connections

In 17 and 18, draw a square with the given area on dot paper.

17. 13 square units

18. 29 square units

19. Which labeled point is the same distance from point A as point B is from point A? Explain.

Extensions

20. Any tilted segment that connects two dots on a sheet of dot paper can be the hypotenuse of a right triangle. You can use this idea to draw segments of a given length. The key is finding two square numbers with a sum equal to the square of the length you want to draw.

For example, to draw a line segment with length $\sqrt{5}$, you can draw a right triangle in which the sum of the areas of the squares on the legs is 5. Since the area of the square on the hypotenuse will then be 5, the length of the hypotenuse will be $\sqrt{5}$.

19. The square of the distance between points A and B is $4^2 + 3^2 = 25$, so the distance between them is $\sqrt{25} = 5$ units. Point F is the other point that is 5 units from point A.

Connections

17.

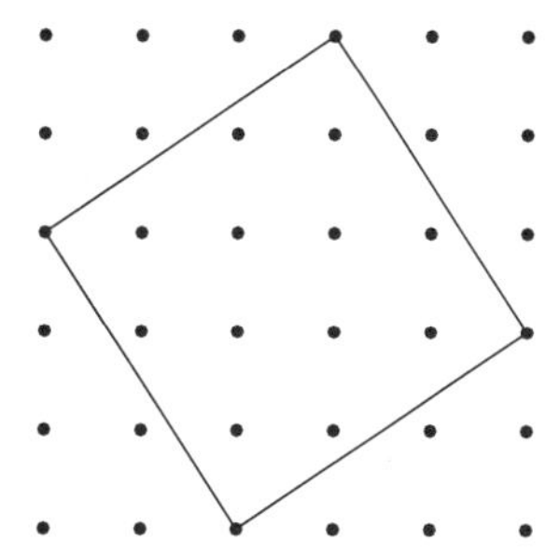

18.

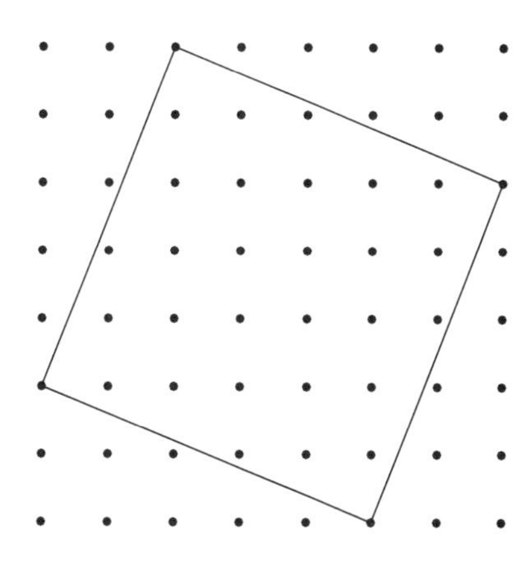

Extensions

20. See page 40m.

21. No; $2^2 = 4$, and 4 can't be represented as the sum of two squares.

22. No; $3^2 = 9$, and 9 can't be represented as the sum of two squares.

23. No; $4^2 = 16$, and 16 can't be represented as the sum of two squares.

24. Yes; $5^2 = 25$, and 25 can be represented as the sum of two squares, 9 + 16.

25. No; $6^2 = 36$, and 36 can't be represented as the sum of two squares.

26. No; $7^2 = 49$, and 49 can't be represented as the sum of two squares.

Since 1 and 4 are square numbers, and 1 + 4 = 5, you can draw a right triangle with legs of lengths 1 and 2.

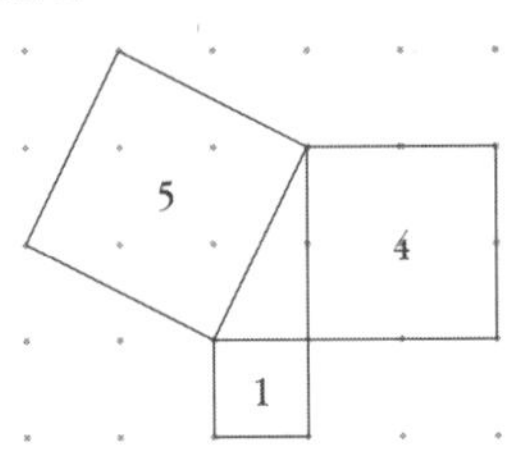

a. To use this technique, it helps to be familiar with sums of square numbers. Copy and complete this table to show the sums of pairs of square numbers.

+	1	4	9	16	25	36	49	64
1	2	5						
4	5							
9								
16								
25								
36								
49								
64								

b. Find two square numbers with the given sum.

i. 10 **ii.** 25 **iii.** 89

c. Use your table to help you draw segments on dot paper with the given length. Explain your work.

i. $\sqrt{26}$ **ii.** 10 **iii.** $\sqrt{50}$

In 21–26, tell whether it is possible to draw a *tilted* segment of the given length by connecting points on dot paper. Explain your reasoning.

21. 2 **22.** 3 **23.** 4

24. 5 **25.** 6 **26.** 7

27. When building a barn, a farmer must make sure the sides of the barn are *perpendicular* to the ground. This means that the sides of the barn form right angles with the ground.

a. One method for checking whether a wall is perpendicular to the ground involves a 10-foot pole. The farmer makes a mark exactly 8 feet high on the wall and then makes a triangle by placing one end of the 10-foot pole on the mark and the other end on the ground.

If the wall is perpendicular to the ground, how far from the base of the wall will the pole touch the ground? Explain how you found your answer.

b. You may have heard the saying, "I wouldn't touch that with a 10-foot pole!" What would this saying mean to a farmer who had just built a barn?

c. Suppose a farmer used a 15-foot pole and made a mark 12 feet high on the wall. If the wall is perpendicular to the ground, how far from the base of the wall will the pole touch the ground?

27a. Assuming the pole is positioned straight out from the wall, the pole must touch the ground 6 feet from the base of the wall because $10^2 - 8^2 = 36$, and $\sqrt{36} = 6$.

27b. Possible answer: A 10-foot pole that couldn't touch the 8-foot mark when placed 6 feet from the base of the wall would mean that the wall was not perpendicular.

27c. The pole must touch the ground 9 feet from the base of the wall because $15^2 - 12^2 = 81$, and $\sqrt{81} = 9$.

Possible Answers

1a. The length of the hypotenuse can be found by taking the square root of the sum of the squares of the lengths of the legs.

1b. Subtract the square of the given leg length from the square of the hypotenuse length; this is the square of the missing leg length. Take the square root of that difference to get the missing leg length.

2. Think of the line segment drawn between the two points as the hypotenuse of a right triangle. Then, find the lengths of the legs of the right triangle (which lie on a vertical line and a horizontal line). Apply the Pythagorean Theorem by adding the squares of these two lengths and taking the square root of that sum.

3. A triangle is a right triangle if the lengths of its sides satisfy the Pythagorean Theorem.

Mathematical Reflections

In this investigation, you worked with a very important mathematical relationship called the Pythagorean Theorem. These questions will help you summarize what you have learned:

1. **a.** Suppose you are given the lengths of the legs of a right triangle. Describe how you can find the length of the hypotenuse.

   b. Suppose you are given the lengths of one leg and the hypotenuse of a right triangle. Describe how you can find the length of the other leg.

2. Describe how you can use the Pythagorean Theorem to find the distance between two dots on a sheet of dot paper without measuring.

3. How can you determine whether a triangle is a right triangle if you know only the lengths of its three sides?

Think about your answers to these questions, discuss your ideas with other students and your teacher, and then write a summary of your findings in your journal.

Tips for the Linguistically Diverse Classroom

Diagram Code The Diagram Code technique is described in detail in *Getting to Know Connected Mathematics*. Students use a minimal number of words and drawings, diagrams, or symbols to respond to questions that require writing. Example: Question 1a—A student might answer this question by drawing a right triangle with an arrow pointing to the hypotenuse and squares drawn on the legs, and labeling the sides of one square *A* and the sides of the other square *B*. Below this, the student might write $\sqrt{A^2 + B^2}$ = hypotenuse.

TEACHING THE INVESTIGATION

3.1 • Discovering the Pythagorean Theorem

In this problem, students collect information about the areas of the squares on the sides of a right triangle. They use the pattern they see in their data to conjecture that the sum of the areas of the two smaller squares equals the area of the largest square.

Launch

To introduce the topic, draw an oblique, or tilted, line segment on a dot grid at the board or overhead.

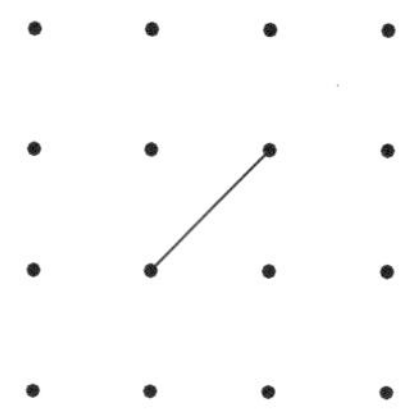

Ask:

> How can we find the length of this line segment? *(We can draw a square using this segment as a side. Then, we can find the area of the square and take its square root.)*

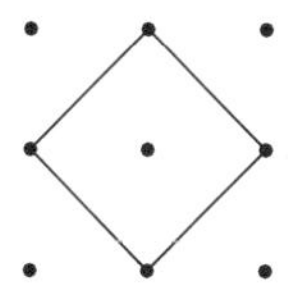

area = 2 square units
segment length = $\sqrt{2}$

Using the original line segment as a hypotenuse, draw two line segments to make a right triangle.

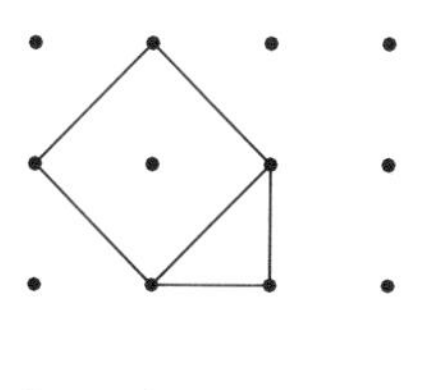

> What kind of triangle have I drawn? *(a right triangle)*

Explain that in a right triangle, the two sides that form the right angle are called the *legs* of the right triangle. The side opposite the right angle is called the *hypotenuse*.

What are the lengths of the two legs of this triangle? What are the squares on the legs? *(In this example, the lengths and areas are both 1.)*

What is the area of the square on the hypotenuse? *(2 square units)*

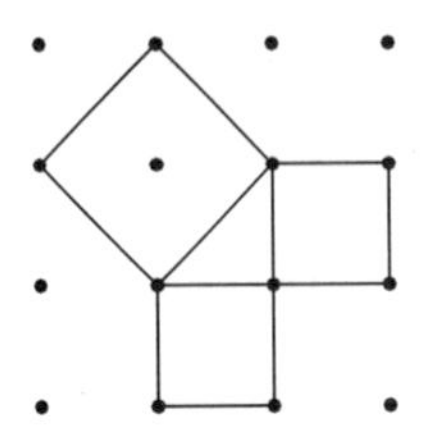

Students may notice that the sum of the area of the squares on the legs is equal to the area of the square on the hypotenuse, but don't push for this observation at this time.

In Problem 3.1, you will be looking for a relationship among the three squares that can be drawn on the sides of a right triangle. It will help to organize your work in a table so that you can look for patterns.

Have students work in groups of three or four on the problem and follow-up.

Explore

Ask that each student complete a table. Encourage the students in each group to share the work, with each student finding the areas for two or three of the right triangles.

As you circulate, check to see that students are correctly drawing the squares on the right triangles.

Summarize

Ask the class to discuss the patterns they see in the table. They should notice that the sum of the areas of the squares on the legs is equal to the area of the square on the hypotenuse.

What conjecture can you make about your results? *(When you add the areas of the squares on the legs, you get the area of the square on the hypotenuse.)* This pattern is called the *Pythagorean Theorem.*

Suppose a right triangle has legs of lengths a and b and a hypotenuse of length c. *(Draw this on the board or overhead.)*

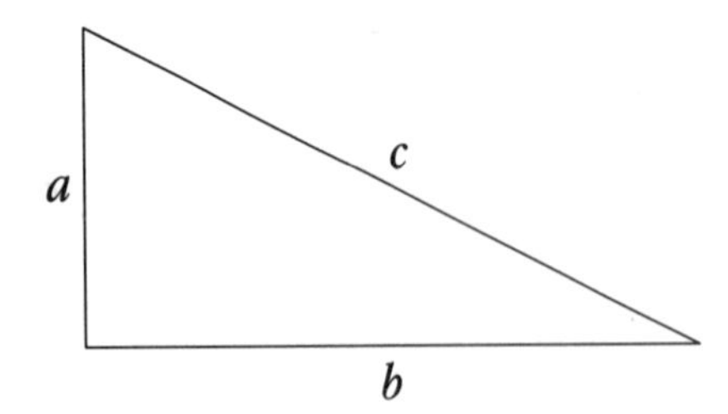

Using these letters, can you state the Pythagorean Theorem in a general way? *(If a and b are the lengths of the legs of a right triangle and c is the length of the hypotenuse, then* $a^2 + b^2 = c^2$*.)*

Do you think the Pythagorean Theorem will work for triangles that are *not* right triangles?

To help the class explore this question, draw the triangle shown below on the board or overhead. Use a corner of a sheet of paper to verify that the triangle does not contain a right angle. Then, draw squares on each side of the triangle.

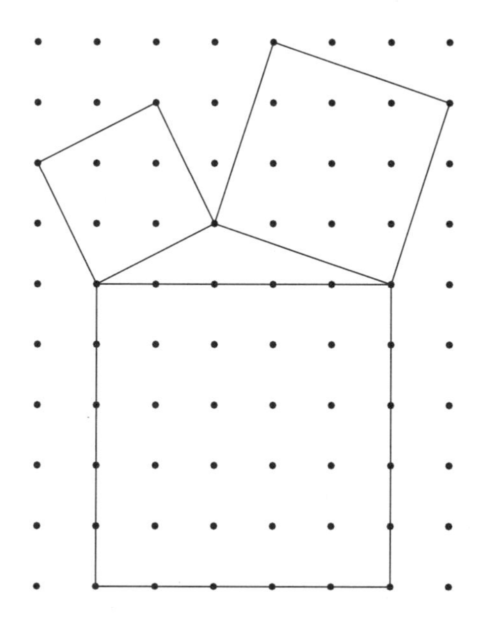

We have shown that this triangle is not a right triangle. What are the areas of the squares on its sides? *(5, 10, and 25 square units)* Do the areas of the squares on the shorter sides equal the area of the square on the longest side? *(no;* $5 + 10 \neq 25$*)*

Next, ask the class this question:

Do you think the Pythagorean Theorem is true for all right triangles, even if the sides are not whole numbers?

The squares on the sides of *any* right triangle are related to each other in such a way that the sum of the areas of the two smaller squares equals the area of the largest square. To help the class explore this, you may want to do ACE question 13 together. This right triangle has sides of lengths $\sqrt{5}$, $\sqrt{5}$, and $\sqrt{10}$. The squares of the side lengths are 5, 5, and 10. Since $5 + 5 = 10$, the Pythagorean Theorem applies to right triangles with side lengths that are not whole numbers.

Go over the follow-up questions. Ask:

How did you find the hypotenuse lengths?

A few students may have used the Pythagorean Theorem; most will have taken the square root of the area of the square on the hypotenuse. Ask for decimal approximations for some of the hypotenuse lengths.

Choose one of the right triangles, list the lengths of the three sides, and ask students what the Pythagorean Theorem says about these lengths.

> The lengths of the sides of a right triangle are 2, 3 and $\sqrt{13}$. What does the Pythagorean Theorem say about these lengths? *[$2^2 + 3^2 = (\sqrt{13})^2$, or $4 + 9 = 13$]*

Repeat the question for lengths 5, 12, and 13.

3.2 • Puzzling Through a Proof

In this problem, students investigate a puzzle that verifies that the sum of the areas of the squares on the legs of a right triangle is equal to the area of the square on the hypotenuse. Students are again introduced to this idea in symbolic form: If a and b are the lengths of the legs of a right triangle and c is the length of the hypotenuse, then $a^2 + b^2 = c^2$.

Launch

Explain to the class that there are many proofs of the Pythagorean Theorem. One of the proofs is based on the puzzle that they will explore in this problem.

Display a set of puzzle pieces on the overhead. Ask students if they see any relationships among the puzzle pieces. Some may notice that the square pieces fit on the sides of the right triangle.

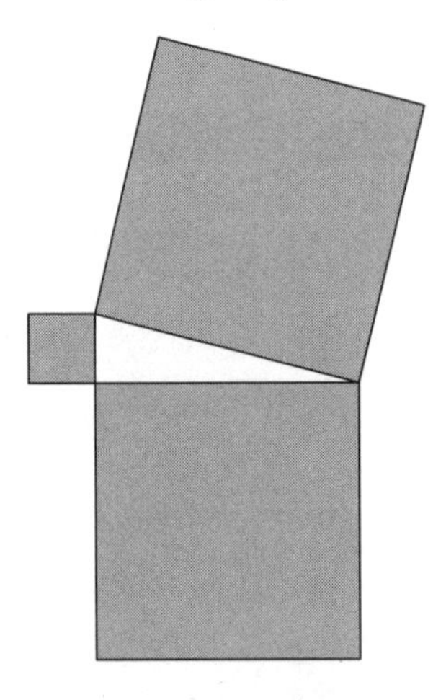

> I'm handing out sheets containing two puzzle frames and 11 puzzle pieces. Your task is to arrange the puzzle pieces in the two frames and to look for a relationship among the areas of the three puzzle squares.

Have students work in groups of four on the problem. Distribute scissors and one copy of Labsheet 3.2 to each student. There are three versions of Labsheet 3.2 (sets A, B, and C); distribute them in approximately equal number, giving all students in a group the same version.

Save the follow-up questions until after the summary of the problem.

Explore

Have each group work with one of the three versions of the puzzle; they will compare their results with another group's results when they have finished the problem.

Encourage each group to find more than one way to fit the puzzle pieces into the two frames.

Summarize

When groups have finished the problem, ask about any general patterns they noticed. Some will mention the relationship between the squares and the sides of the right triangle; some may notice that a side length of a puzzle frame is equal to the sum of the lengths of the two legs of each triangle. Demonstrate these relationships at the overhead.

Have a couple of groups show how they arranged their puzzle pieces. The arrangements may differ slightly, but they all lead to the same conclusion. One arrangement is shown below.

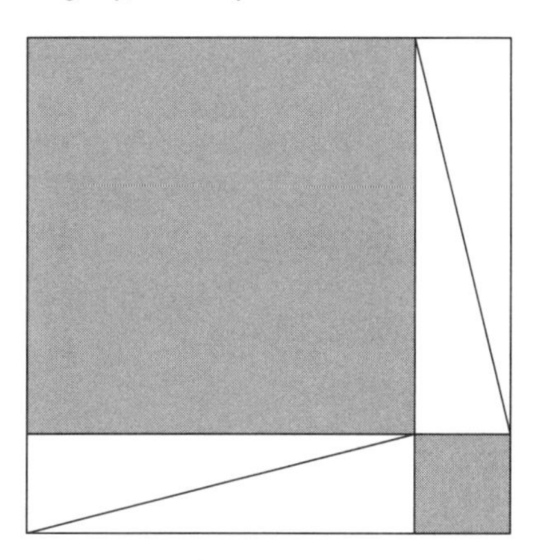

What relationship do these completed puzzles suggest?

Use follow-up question 1 as a whole-class discussion to generalize and to help students to understand the following argument: The areas of the frames are equal. Each frame contains four identical right triangles. If the four right triangles are removed from each frame, the area remaining in the frames must be equal. That is, the sum of the areas of the squares in one frame must equal the area of the square in the other frame.

Label one of the arrangements suggested by the class as shown below.

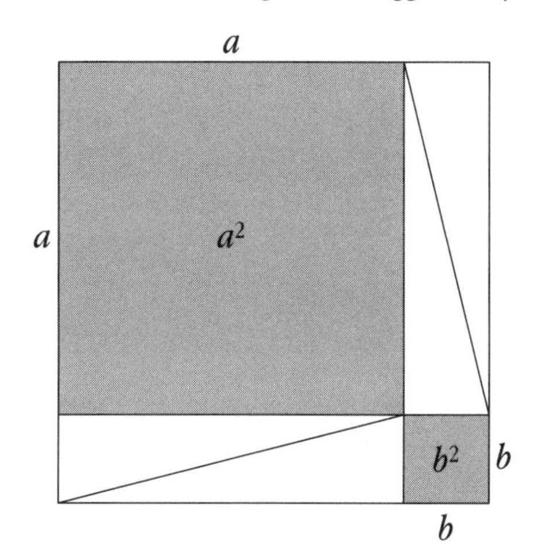

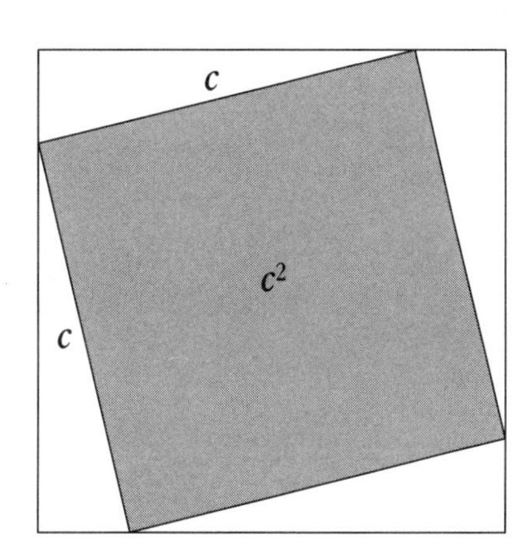

The diagram shows that if the lengths of the legs of a right triangle are a and b and the length of the hypotenuse is c, then $a^2 + b^2 = c^2$. This is a geometric proof of the Pythagorean Theorem.

Before having them consider the follow-up questions, offer an example to help them apply the theorem.

> How can you use the Pythagorean Theorem to find the length of the hypotenuse of a right triangle? *(If we know the lengths of the legs, we can find the areas of the squares on those two sides and add them. This total area is equal to the area of the square on the hypotenuse. Taking the square root of that amount will give us the length of the hypotenuse.)*

Draw these triangles on the board or overhead:

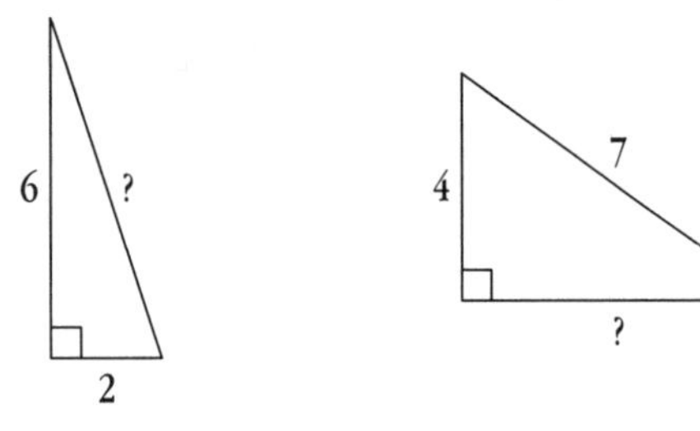

> How might we find the missing side lengths in these right triangles?

Students will likely suggest finding the areas of the squares on the labeled sides. For the triangle on the left, the areas of the squares on the legs are 36 and 4. The sum, 40, is the area of the square on the hypotenuse; the length of the hypotenuse is thus $\sqrt{40}$. For the triangle on the right, the area of the square on the hypotenuse is 49, which is equal to the sum of the areas of the squares on the legs. The area of the square on the unlabeled leg is thus $49 - 16 = 33$; the length of the missing leg must be $\sqrt{33}$.

Assign follow-up questions 2 to 5 as a check of students' understanding of the Pythagorean Theorem.

3.3 • Finding Distances

In this problem, students discover how the Pythagorean Theorem can be used to find the distance between two dots on a grid.

Launch

Display Transparency 3.3 or a transparent grid, and indicate or label points A and B as shown in the student edition. Ask:

> How can you find the distance between these two points?

For the Teacher: Geometry, Algebra, and the Pythagroean Theorem

The Pythagorean Theorem describes a relationship among the lengths of the sides of a right triangle. This relationship is regarded as one of the most important developments in mathematics as it allows us to link ideas of number to ideas of space. We can assign a number to the length of a line segment. The scheme of assigning a number to the length of a line segment was developed by René Descartes in 1637 when he created the coordinate plane.

The geometric argument explored in the Summarize section can be considered algebraically.

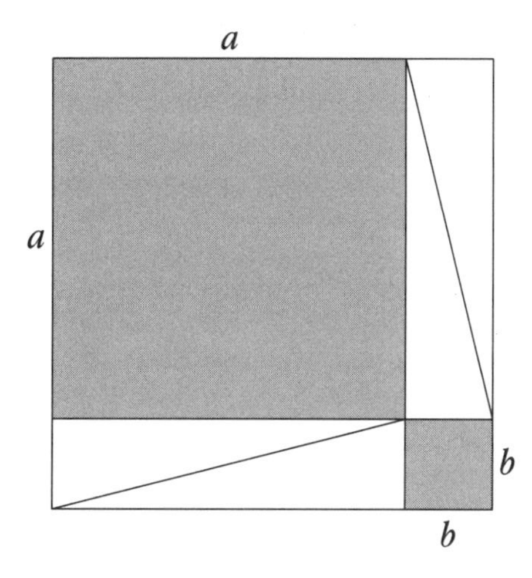

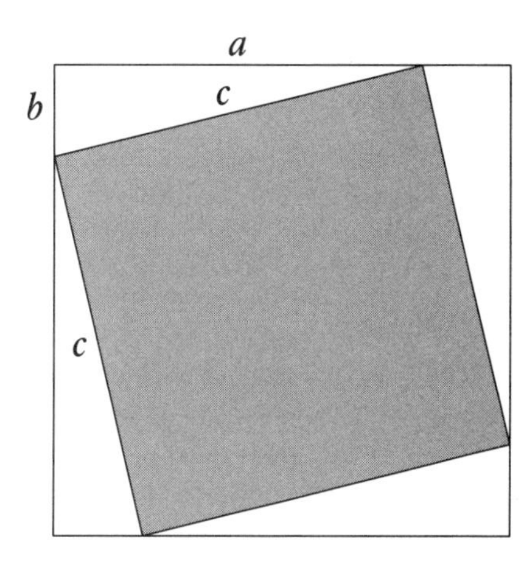

The sum of the areas of the two squares and the four triangles in the left frame equals the sum of the areas of the square and the four triangles in the right frame:

$$a^2 + b^2 + 4(\tfrac{ab}{2}) = c^2 + 4(\tfrac{ab}{2})$$
$$a^2 + b^2 = c^2$$

The class may suggest measuring the distance with a ruler. Explain that the Pythagorean Theorem can be used to find a more exact length. Use points *A* and *B* to illustrate the method of finding the distance. Draw line segment *AB* and ask:

How can we use the Pythagorean Theorem to find the length of this line segment?

Some students will probably suggest using the segment as the side of a square; others may suggest using it as the hypotenuse of a right triangle.

What right triangle has this hypotenuse?

Sketch students' suggestions, which may be either of the triangles shown here:

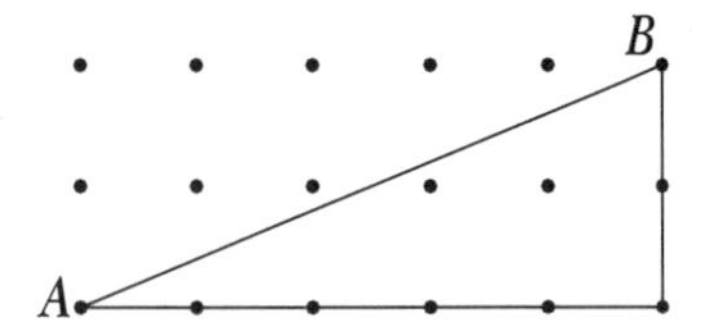

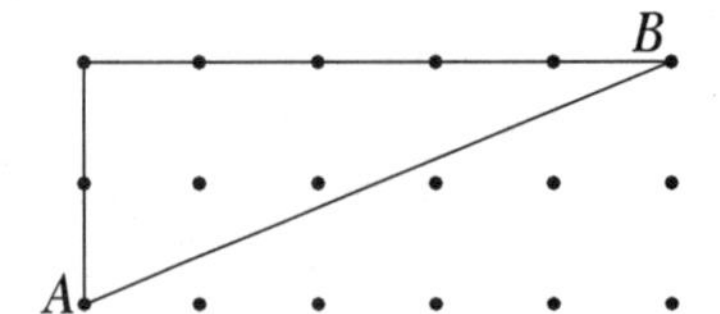

What are the lengths of the legs? *(5 and 2)* How can you use this information to find the length of the hypotenuse? *(The square of the length of the hypotenuse = $5^2 + 2^2 = 29$, so the length is $\sqrt{29}$.)* So, what is the distance between points *A* and *B*? *($\sqrt{29}$)*

Distribute Labsheet 3.3 to each student, and have the class work in pairs on the rest of the problem and the follow-up.

Explore

Students should find the problem a review of what they have learned so far. However, the follow-up is a bit difficult, so you may need to help guide their thinking.

Can the $\sqrt{13}$-unit line segment be a vertical or a horizontal segment? *(no)* Why not?

If it is a tilted line segment, can it be the hypotenuse of a right triangle? *(yes)*

Assume that this segment is the hypotenuse of a right triangle. What will the area of the square on the hypotenuse be? *[$(\sqrt{13})^2$, or 13, square units]*

What is the sum of the areas of the squares on the legs of this right triangle? *(13)*

What are two squares whose sum is 13? *(4 and 9)* So, what are the lengths of the legs? *($\sqrt{4}$ and $\sqrt{9}$, or 2 and 3)*

Now, draw a right triangle with legs of lengths 2 and 3. The hypotenuse has length $\sqrt{13}$.

Summarize

Ask students to demonstrate and explain how they found the answers to the problem.

Then, go over the follow-up carefully. After someone has explained how he or she found two points that were $\sqrt{13}$ units apart, offer a similar problem.

How would you find a line segment with a length of $\sqrt{40}$?

Ask one or two students to describe their method. They will likely have used a guess-and-check procedure to find the two squares with a sum of 40, which are 36 and 4. From this they can determine that the lengths 6 and 2 will give a right triangle with a hypotenuse of length $\sqrt{40}$.

Students should be able to focus on the areas of the three squares on the sides of a right triangle and their relationship to the lengths of the sides. Typically, two lengths or two areas are known, and we must find the third length or area. Once we know the missing area, we can take its square root to find the length. Conversely, once we know the missing length, we can square it to find the area.

The following visual explanation will help some students to understand the essence of the Pythagorean Theorem:

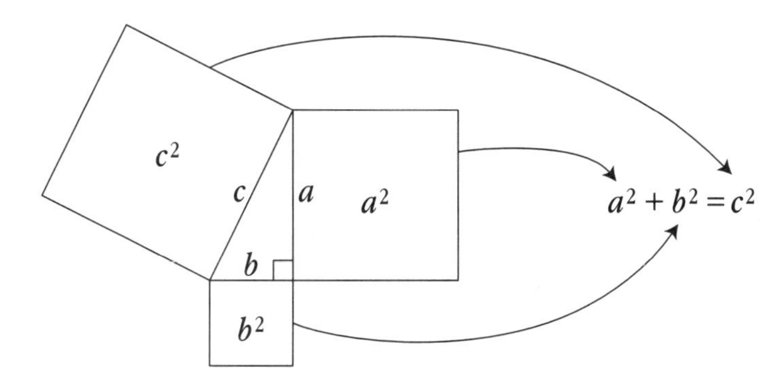

The essential strategy for finding a tilted line with a certain length depends on finding two squares whose sum is equal to the square of that length. If students have done ACE question 20, the table of squares they created can help them to find the sum of the areas of upright squares on the legs of a right triangle to create a hypotenuse with the desired length. As a final check, ask this question:

> Can 7 be the length of a tilted line segment drawn between two dots on a dot grid? *(No, because 49 does not equal the sum of two squares of whole numbers.)*

3.4 • Measuring the Egyptian Way

In this problem, students investigate the converse of the Pythagorean Theorem: If a, b, and c are the lengths of the sides of a triangle and $a^2 + b^2 = c^2$, then the triangle is a right triangle.

Launch

Talk about the two questions in the introduction to Problem 3.4:

- Is any triangle whose side lengths satisfy the relationship $a^2 + b^2 = c^2$ a right triangle?
- If the side lengths of a triangle do not satisfy the relationship $a^2 + b^2 = c^2$, does this mean that the triangle is not a right triangle?

Distribute rulers and straws or string, and have the class work in pairs on the problem and follow-up.

Explore

If necessary, help students form a triangle for parts A and B.

Challenge some students to think about the multiples of side lengths of 3-4-5 and 5-12-13, such as 6-8-10 and 10-24-26.

> Do triangles whose sides have these lengths form a right triangle as well? How do you know? *(yes, because $6^2 + 8^2 = 10^2$ and $10^2 + 24^2 = 26^2$)*

Summarize

Have someone demonstrate at the overhead how he or she arranged string or straws to form a 3-4-5 right triangle. Ask the student how he or she knew it was a right triangle.

Demonstrate, or have a student demonstrate, how the Egyptians used this set of numbers as lengths along a rope to measure right angles (this idea is addressed in the follow-up). By thinking of the knots as knots 0 through 12; holding knots 0, 3, and 7; and putting knot 12 with knot 0, you will make a right triangle.

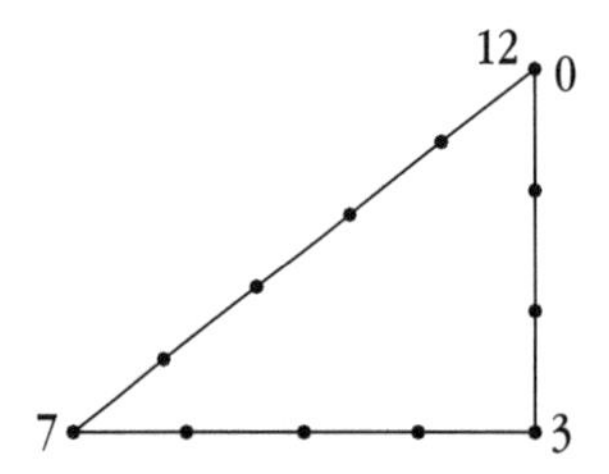

Ask the class whether multiples of a 3-4-5 triangle, such as 6-8-10 and 30-40-50 triangles, will also form right triangles. Have the class check these.

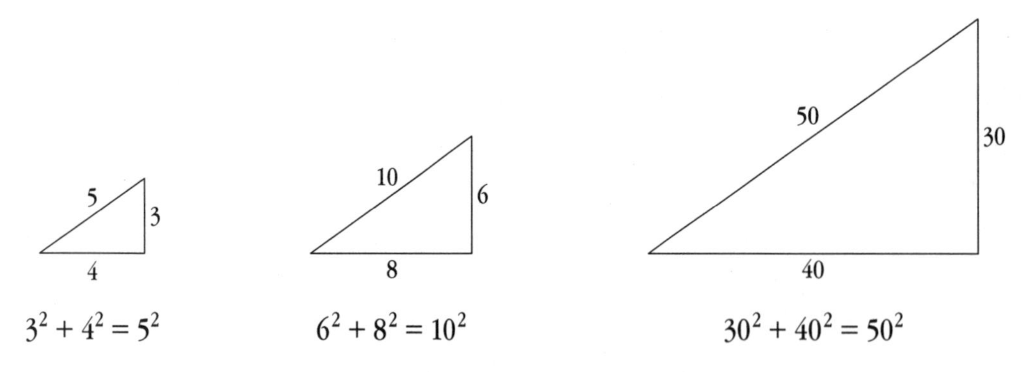

$3^2 + 4^2 = 5^2$ $\qquad$ $6^2 + 8^2 = 10^2$ $\qquad$ $30^2 + 40^2 = 50^2$

> What about the multiples of 5-12-13? Do these lengths form a right triangle? *(yes, $10^2 + 24^2 = 26^2$, $15^2 + 36^2 = 39^2$, and so on)*
>
> Set of three numbers that satisfy the Pythagorean Theorem are called *Pythagorean triples.* Other whole-number triples are 7-24-25 and 9-40-41.

Go over the follow-up questions. The triangle in question 1, part c, is not a right triangle. Have students demonstrate each set of lengths on a grid at the overhead, checking for right angles with an angle ruler or a corner of a piece of paper.

Additional Answers

Answers to Problem 3.1

A, B.

Length of leg 1	Length of leg 2	Area of square on leg 1	Area of square on leg 2	Area of square on hypotenuse
1	1	1	1	2
1	2	1	4	5
2	2	4	4	8
1	3	1	9	10
2	3	4	9	13
3	3	9	9	18
3	4	9	16	25

C. The area of the square on the hypotenuse is equal to the sum of the areas of the squares on the legs.

D. Possible answer: If the legs of a right triangle are 4 and 1, then the area of the square on the hypotenuse is 17, since 16 + 1 = 17.

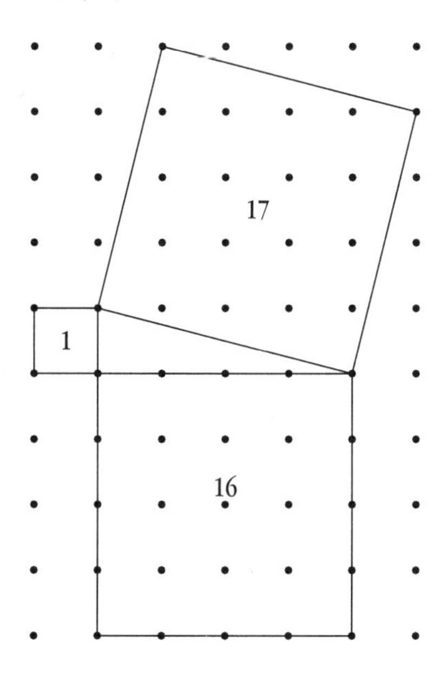

Answers to Problem 3.1 Follow-Up

1, 2.

Length of hypotenuse	Approximate length
$\sqrt{2}$	1.41
$\sqrt{5}$	2.24
$\sqrt{8}$	2.83
$\sqrt{10}$	3.16
$\sqrt{13}$	3.61
$\sqrt{18}$	4.24
$\sqrt{25}$	5.00

Answers to Problem 3.3

A. 1.

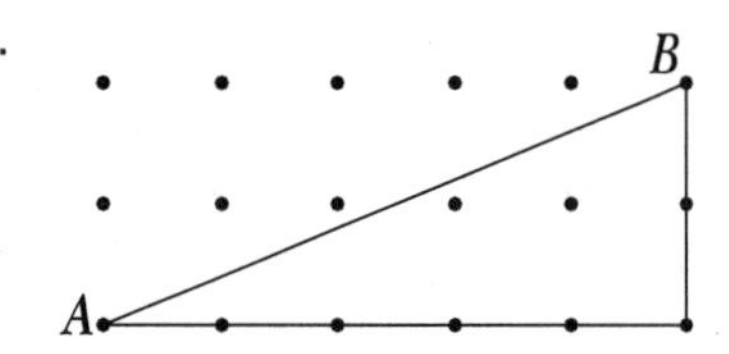

2. 2 and 5

3. Since $5^2 + 2^2 = 29$ square units, the hypotenuse has a length of $\sqrt{29}$.

B. Since $4^2 + 3^2 = 25$ square units, the hypotenuse has a length of $\sqrt{25} = 5$.

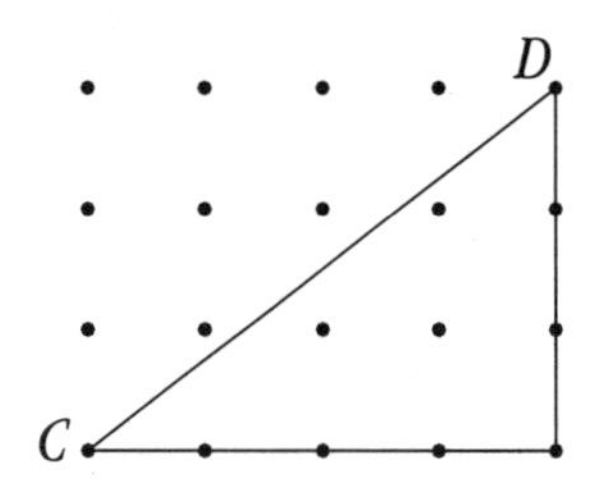

C. Since $6^2 + 3^2 = 45$ square units, the hypotenuse has a length of $\sqrt{45}$.

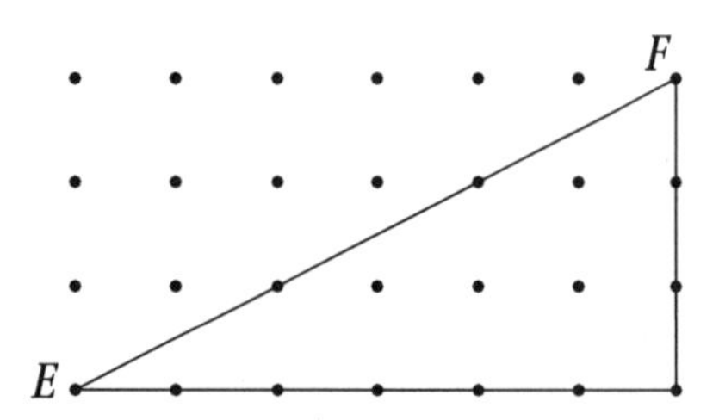

Answers to Problem 3.4

A. 1. Yes; the values $a = 3$, $b = 4$, and $c = 5$ satisfy the relationship, as $3^2 + 4^2 = 5^2$.

2.

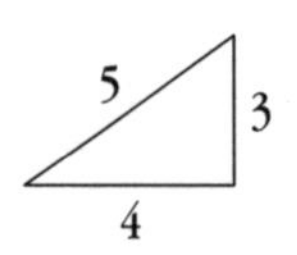

3. This is a right triangle.

4. $5^2 + 12^2 = 13^2$

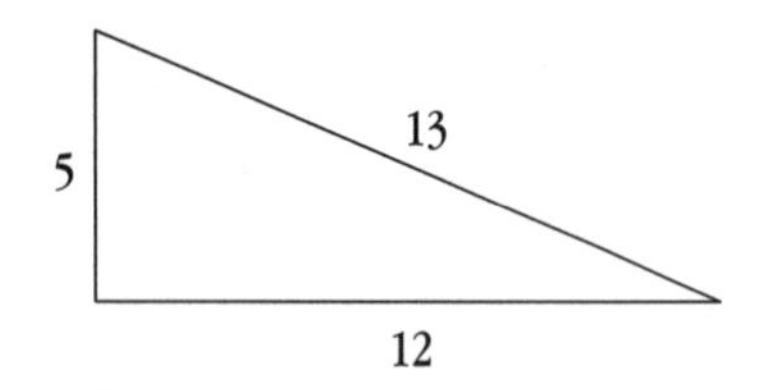

5. If a triangle's side lengths satisfy the relationship $a^2 + b^2 = c^2$, the triangle is a right triangle.

B. 1. Possible answer: A triangle with $a = 1$, $b = 1$, and $c = 1$ does not satisfy the relationship $a^2 + b^2 = c^2$.

2. This is not a right triangle.

3. A triangle with $a = 2$, $b = 2$, and $c = 3$ does not satisfy the relationship $a^2 + b^2 = c^2$.

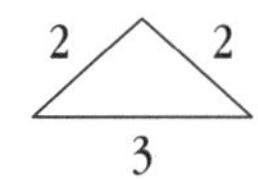

4. If a triangle's side lengths do not satisfy the relationship $a^2 + b^2 = c^2$, the triangle is not a right triangle.

ACE Answers

Applications

13. The squares on the legs each have an area of 5 square units, giving a sum of 10 square units—which is the same as the area of the square on the hypotenuse.

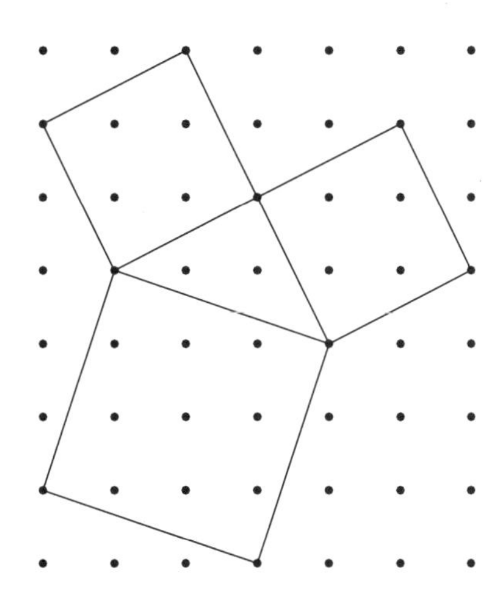

20a.

+	1	4	9	16	25	36	49	64
1	2	5	10	17	26	37	50	65
4	5	8	13	20	29	40	53	68
9	10	13	18	25	34	45	58	73
16	17	20	25	32	41	52	65	80
25	26	29	34	41	50	61	74	89
36	37	40	45	52	61	72	85	100
49	50	53	58	65	74	85	98	113
64	65	68	73	80	89	100	113	128

20b. **i.** 1 and 9 **ii.** 9 and 16 **iii.** 25 and 64

20c. Square the length and find that number in the chart. The numbers in the heads of that column and row indicate the areas of the squares needed on the legs of a right triangle that has a hypotenuse of that length. (**Teaching Tip:** You may want to suggest that students use the table from this question to help solve other problems.)

i.

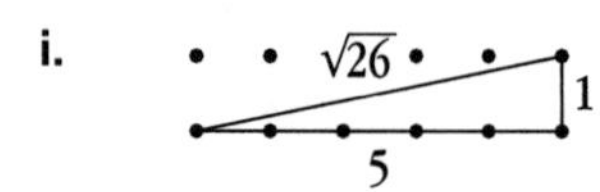

ii.

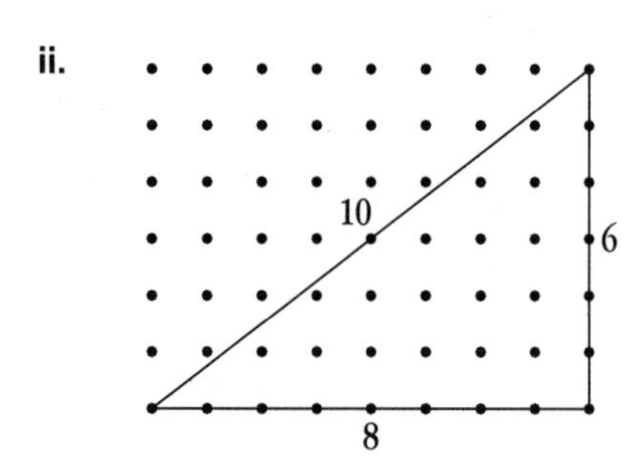

iii.

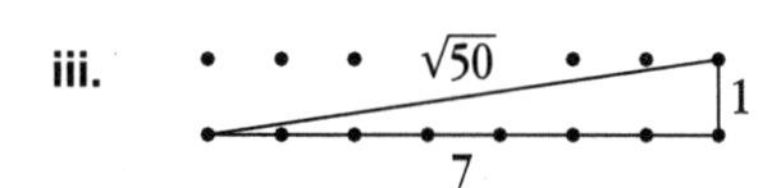

Using the Pythagorean Theorem

For students to begin to appreciate the mathematical power of the Pythagorean Theorem, they need to encounter situations that can be illuminated by the theorem. This investigation provides those experiences.

In Problem 4.1, Stopping Sneaky Sally, students apply the Pythagorean Theorem to find distances on a baseball diamond.

In Problem 4.2, Analyzing Triangles, students investigate properties of some special right triangles, including a 30-60-90 triangle and an isosceles right triangle, by applying the Pythagorean Theorem.

In Problem 4.3, Finding the Perimeter, students draw from their experiences in the previous two problems to find missing lengths and angles in a group of triangles.

Mathematical and Problem-Solving Goals

- ***To apply the Pythagorean Theorem in several problem situations***
- ***To investigate the special properties of a 30-60-90 triangle and an isosceles right triangle***
- ***To use the properties of special right triangles to solve problems***

Materials

Problem	For students	For the teacher
All	Graphing calculators	Transparencies: 4.1 to 4.3 (optional)
4.1	Dot paper	Transparent dot paper (optional)
4.2	Labsheet 4.2 (1 per pair), scissors	Transparency of Labsheet 4.2 (optional)

Using the Pythagorean Theorem

The Pythagorean Theorem is an important and useful mathematical idea. In this investigation, you will use the theorem to solve some interesting problems.

The Pythagorean Theorem

If a right triangle has legs of lengths a and b and a hypotenuse of length c, then $a^2 + b^2 = c^2$.

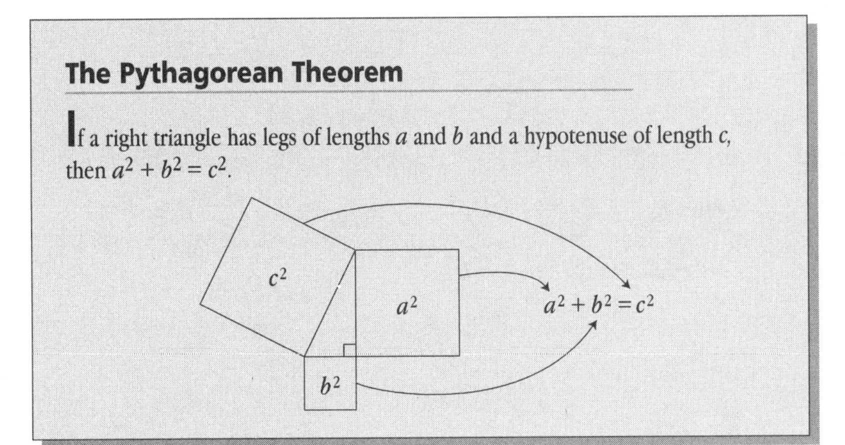

4.1 Stopping Sneaky Sally

The Pythagorean Theorem can be used in situations like the following, in which you need to find a missing length in a right triangle.

Horace Hanson is the catcher for the Humbolt Bees baseball team. Sneaky Sally Smith, the star of the Canfield Cats, is on first base. Sally is known for stealing bases, so Horace is keeping a sharp eye on her.

The pitcher throws a fastball, and the batter swings and misses. Horace catches the pitch. Out of the corner of his eye, he sees Sally take off for second base.

4.1 Stopping Sneaky Sally

At a Glance

Grouping: *pairs*

Launch

- Introduce the baseball game being played by Horace and Sally.
- Talk about the layout of a baseball diamond.
- Have pairs explore the problem and follow-up.

Explore

- Offer help to students who are having trouble applying the Pythagorean Theorem.

Summarize

- Have students share their strategies for finding the two distances.
- Address any misconceptions about applying the theorem and working with square roots.

Tips for the Linguistically Diverse Classroom

Enactment The Enactment technique is described in detail in *Getting to Know Connected Mathematics.* Students act out mini-scenes using props to make information comprehensible. Example: For "Stopping Sneaky Sally," ask students to play the roles of Horace, Sally, the pitcher, and the batter to act out the scenario above. The pitcher pantomimes throwing the ball, the batter swings, and Sally takes off for second base. Horace pantomimes getting ready to throw the ball to second base. Pieces of heavy paper arranged as in a baseball diamond can serve as props.

Assignment Choices

ACE questions 1, 3, 5, 6, and unassigned choices from earlier problems

Problem 4.1

How far must Horace throw the baseball to get Sally out at second base? Explain how you found your answer.

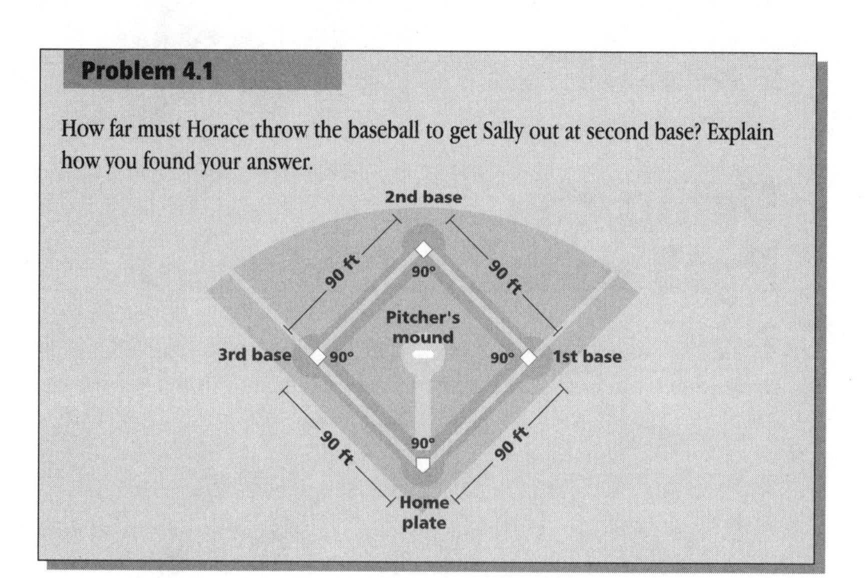

Problem 4.1 Follow-Up

The shortstop is standing on the baseline, halfway between second base and third base. How far is the shortstop from Horace?

Did you know?

Although most people consider baseball an American invention, a similar game, called *rounders,* was played in England as early as the 1600s. Like baseball, rounders involved hitting a ball and running around bases. However, in rounders, the fielders actually threw the ball at the base runners. If a ball hit a runner while he was off base, the runner was out.

Alexander Cartwright is considered the father of organized baseball. He started the Knickerbockers Base Ball Club of New York in 1845 and wrote an official set of rules. According to Cartwright's rules, a batter was out if a fielder caught the ball either on the fly or on the first bounce. Today, balls caught on the first bounce are not outs. Cartwright's rules also stated that the first team to have a total of 21 runs at the end of an inning was the winner. Today, the team with the highest score after nine innings wins the game.

Answer to Problem 4.1

Since $90^2 + 90^2 = 16{,}200$, the distance from home plate to second base is $\sqrt{16{,}200} \approx 127$ ft.

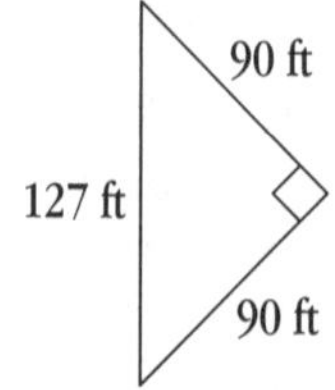

Answer to Problem 4.1 Follow-Up

The shortstop is standing on the baseline at a distance of $90 \div 2 = 45$ ft from third base, and $90^2 + 45^2 = 10{,}125$, so the distance from home plate to the shortstop is $\sqrt{10{,}125} \approx 101$ ft.

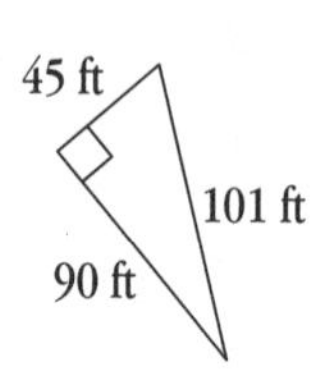

Analyzing Triangles

In this problem, you will work with an equilateral triangle. An *equilateral triangle* is a triangle in which all three sides are the same length. What is true about the angle measures in an equilateral triangle?

Problem 4.2

Each side of equilateral triangle ABC has a length of 2.

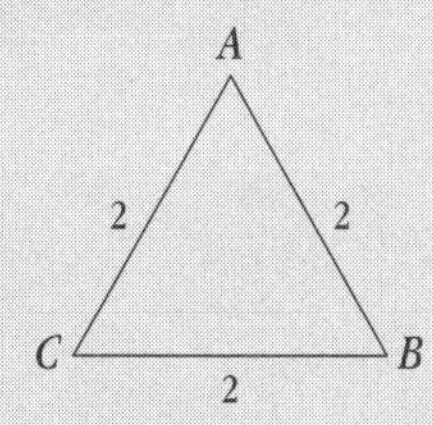

On Labsheet 4.2, find the point halfway between vertices B and C. Label this point P. Point P is the *midpoint* of segment BC. Draw a segment from vertex A to point P. This divides triangle ABC into two triangles. Cut out triangle ABC and fold it along line AP.

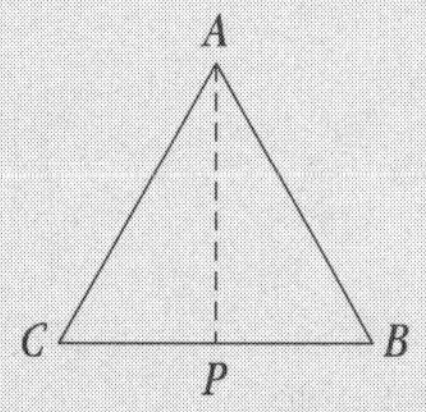

A. How does triangle ABP compare with triangle ACP?

B. Find the measure of each angle in triangle ABP. Explain how you found each measure.

C. Find the length of each side of triangle ABP. Explain how you found each length.

D. Two line segments that meet at right angles are called **perpendicular** line segments. Find a pair of perpendicular line segments in the drawing above.

E. What relationships do you observe among the side lengths of triangle ABP? Are these relationships also true for triangle ACP? Explain.

Analyzing Triangles

At a Glance

Grouping: ***pairs***

Launch

- Draw an equilateral triangle, and discuss its properties with the class.
- Distribute Labsheet 4.2 and scissors, and have pairs work on the problem and follow-up.

Explore

- Have students label their paper triangles.
- Help students to see that two right triangles formed from halving an equilateral triangle have special properties.

Summarize

- Have students share what they learned about the relationships in a 30-60-90 triangle.
- Review the follow-up questions and what students discovered about an isosceles right triangle.

Answers to Problem 4.2

A. Measures of corresponding angles and sides are equal, so triangles ABP and ACP are congruent.

B. See page 52i.

C. The length of side AB is 2 because it is a side of the original equilateral triangle. The length of side BP must be 1, as point P divides side BC in half and the length of side BC is 2. Because triangle APB is a right triangle, and $2^2 - 1^2 = 3$, the length of side AP is $\sqrt{3}$.

D. Line segments BC and AP are perpendicular.

E. The length of the side opposite a 30° angle in a right triangle is half the length of the hypotenuse, and the longer leg is $\sqrt{3}$ times the length of the shorter leg.

Assignment Choices

ACE questions 2, 4, 7, 10, and unassigned choices from earlier problems

Problem 4.2 Follow-Up

1. A right triangle with a 60° angle is sometimes called a *30-60-90 triangle.* This 30-60-90 triangle has a hypotenuse of length 6. What are the lengths of the other two sides? Explain how you found your answers.

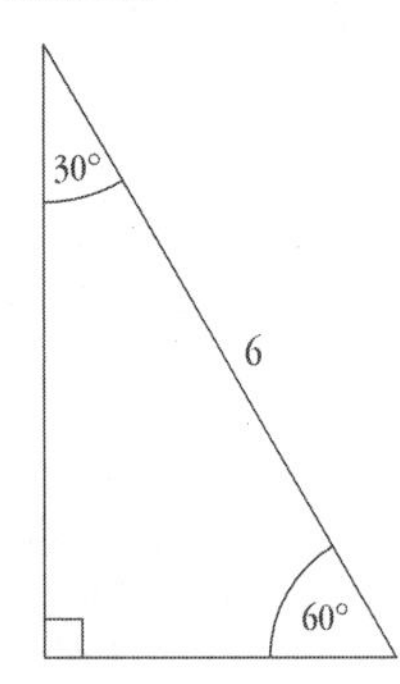

2. Square *ABCD* has sides of length 1. On Labsheet 4.2, draw a diagonal, dividing the square into two triangles. Cut out the square and fold it along the diagonal.

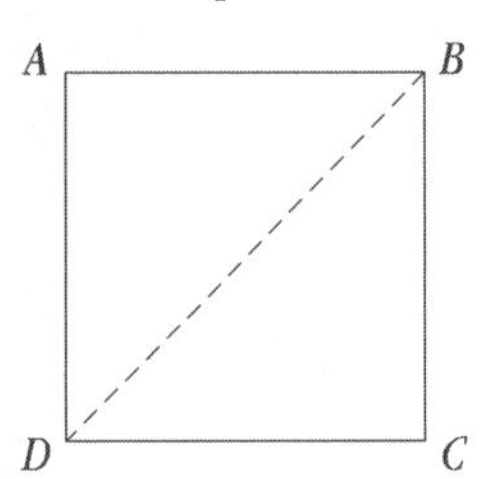

a. How do the two triangles compare?

b. What are the measures of the angles of one of the triangles? Explain how you found each measure.

c. What is the length of the diagonal? Explain how you found the length.

d. Suppose square *ABCD* had sides of length 5. How would this change your answers to parts b and c?

Answers to Problem 4.2 Follow-Up

1. See page 52i.

2. a. The triangles have the same shape and size; measures of corresponding angles and sides are equal. They are congruent.

   b. The angle measures in each triangle are 45°, 45°, and 90°. The diagonal line divides the corner angles into two equal angles, so the smaller angles must each be half of 90°, or 45°.

   c. The legs of the right triangle each have a length of 1, and $1^2 + 1^2 = 2$, so the diagonal—which is the hypotenuse—has a length of $\sqrt{2}$.

   d. The measures of the angles would still be 45°, 45°, and 90°. Since $5^2 + 5^2 = 50$, the length of the diagonal would be $\sqrt{50}$. (Note: Some students may notice that $\sqrt{50} = \sqrt{25 \times 2} = 5\sqrt{2}$.)

4.3 Finding the Perimeter

In this problem, you will apply many of the strategies you have developed in this unit.

Problem 4.3

In the diagram below, some lengths and angle measures are given. Use this information and what you have learned in this unit to help you find the perimeter of triangle *ABC*. Explain your work.

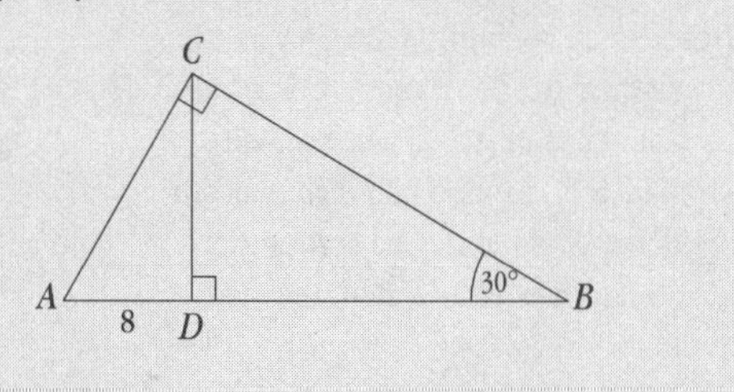

Problem 4.3 Follow-Up

1. Find the area of triangle *ABC*. Explain your reasoning.
2. Find the areas of triangle *ACD* and triangle *BCD*.

Finding the Perimeter

At a Glance

Grouping: *small groups*

Launch

- Discuss triangle *ABC* as shown in the student edition and what is necessary to find its perimeter.
- Have groups of four work on the problem and follow-up.

Explore

- Help students who have trouble seeing the three 30-60-90 triangles.
- Remind students to keep track of their calculations.

Summarize

- Have one group explain how they found the perimeter.
- Discuss how students found the areas of the triangles in the follow-up.

Answer to Problem 4.3

Side *AC* has a length of 16, side *AB* has a length of 32, and side *BC* has a length of $\sqrt{768}$. The perimeter is thus $16 + 32 + \sqrt{768} \approx 75.7$. Explanations will vary; see the possible explanation in the Summarize section.

Answers to Problem 4.3 Follow-Up

1. The area of triangle *ABC* is $\frac{1}{2}bh = \frac{1}{2} \times 16 \times \sqrt{768} \approx 221.7$ square units.
2. Using the Pythagorean Theorem, since $16^2 - 8^2 = 192$, the length of side *CD* is $\sqrt{192}$. So, the area of triangle *ACD* is $\frac{1}{2}bh = \frac{1}{2} \times 8 \times \sqrt{192} \approx 55.4$ square units. The length of side *BD* is $32 - 8 = 24$, so the area of triangle *BCD* is $\frac{1}{2} \times 24 \times \sqrt{192} \approx 166.3$ square units.

Assignment Choices

ACE questions 8, 9, 11, 12, and unassigned choices from earlier problems

Assessment

It is appropriate to use the quiz after this problem.

Answers

Applications

1a. Since $500^2 + 600^2 = 610{,}000$, the distance is $\sqrt{610{,}000} \approx 781$ m.

1b. $1100 - 781 = 319$ m

Applications • Connections • Extensions

As you work on these ACE questions, use your calculator whenever you need it.

Applications

1. Scott, a freshman at Michigan State University, needs to walk from his dorm room in Wilson Hall to his math class in Wells Hall. Normally, he walks 500 meters east and 600 meters north along the sidewalks, but today he is running late. He decides to take the shortcut through the Tundra.

a. How many meters long is Scott's shortcut?

b. How much shorter is the shortcut than Scott's usual route?

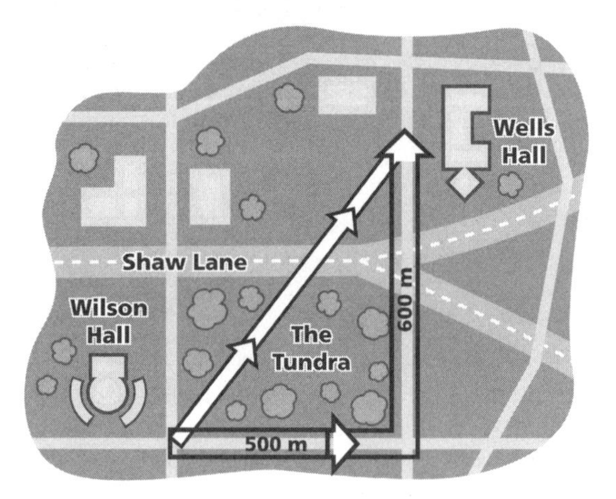

2. Lotsafun Amusement Park has a new ride, the Sky Breaker. The starting and ending points of the ride are separated by 1000 meters. The tram cars glide along a cable that rises at a 45° angle from the ground until it reaches a height of 15 meters. The cable runs parallel to the ground for most of the ride, eventually sloping down again at a 45° angle with the ground. How long, to the nearest tenth of a meter, is the cable for the Sky Breaker ride?

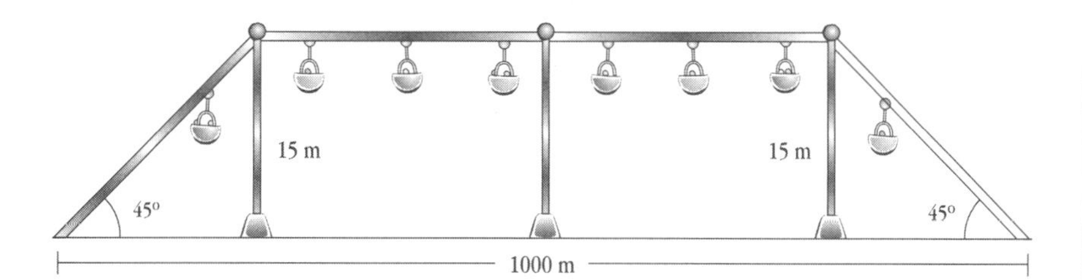

3. At Errol's Evergreen Farm, the taller trees are braced by a wire extending from 2 feet below the top of the tree to a stake in the ground. What is the tallest tree that can be braced with a 25-foot wire staked 15 feet from the base of the tree?

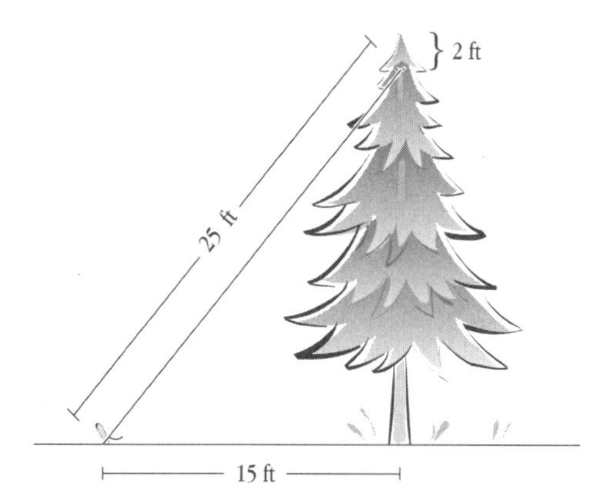

2. The first segment along the ground is the leg of an isosceles right triangle. Since the other leg is 15 m, this leg also has a length of 15 m. The horizontal length of cable is $1000 - (2 \times 15) = 970$ m long. Since $15^2 + 15^2 = 450$, each angled piece of cable (the hypotenuse of the triangle) has a length of $\sqrt{450} \approx 21.2$ m. The overall length of the cable is thus $970 + 21.2 + 21.2 \approx 1012.4$ m.

3. Since $25^2 - 15^2 = 400$, the tallest tree that can be braced is $\sqrt{400} = 20$ ft tall at the point of attachment, plus 2 ft, for a total of 22 ft.

4. The leg along the bottom of the 30-60-90 triangle measures 58 ft. The hypotenuse (from Jeff's eyes to the top of the tower) is twice as long, or 116 ft. Since $116^2 - 58^2 = 10{,}092$, the vertical leg measures $\sqrt{10{,}092} \approx 100.5$ ft. Adding the distance from the ground to Jeff's eyes, Beaumont Tower is about 105.5 ft tall.

Connections

5. The bottom of the box has sides of length 3 cm and 4 cm, and $3^2 + 4^2 = 25$, so the diagonal of the bottom of the box has length $\sqrt{25} = 5$ cm. Using this as a leg of a right triangle with hypotenuse d, $d^2 = 5^2 + 12^2 = 169$, so $d = \sqrt{169} = 13$ cm.

4. As part of his math assignment, Jeff has to estimate the height of Beaumont Tower without measuring it. His class has just studied 30-60-90 triangles, and Jeff decides to use what he learned to help him estimate the height of the tower.

Jeff uses an angle ruler to sight the top of the tower at 60°. Then, he marks the spot where he is standing and measures the distance from the tower to his mark. The distance is about 58 feet. When Jeff stands, his eyes are 5 feet from the ground. About how tall is Beaumont Tower? Explain how you found your answer.

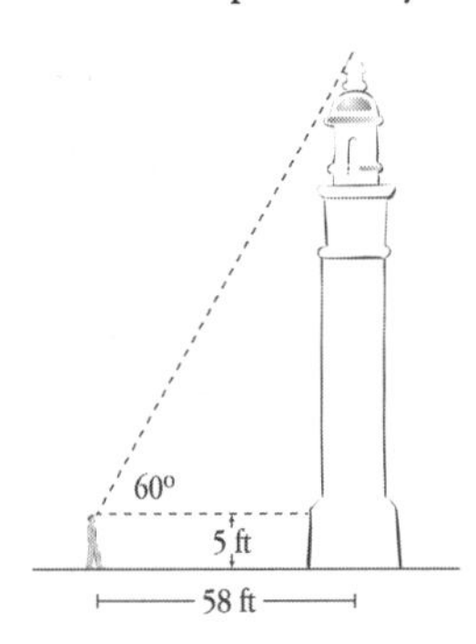

Connections

5. Find the length of the diagonal, d, of the box.

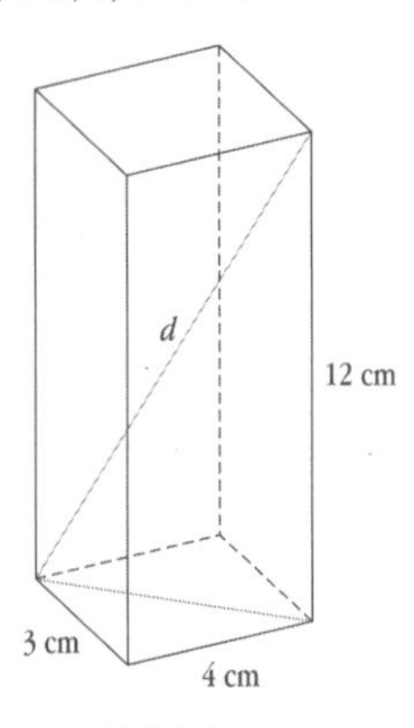

6. Find the length of the diagonal, d, of the box.

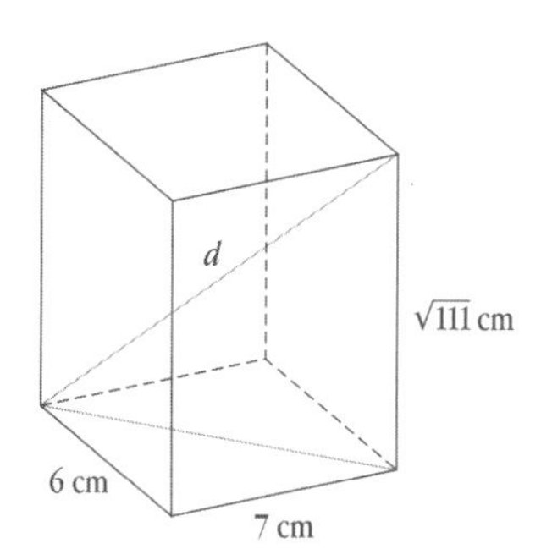

In 7 and 8, use this information: Two cars leave the city of Walleroo at noon. One car travels north and the other travels east.

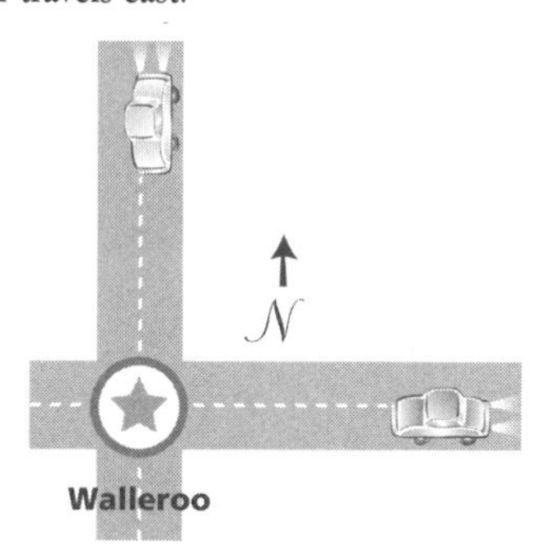

7. Suppose the northbound car is traveling at 60 miles per hour and the eastbound car is traveling at 50 miles per hour. Make a table that shows the distance each car has traveled and the distance between the two cars after 1 hour, 2 hours, 3 hours, and so on. Describe how the distances are changing.

8. Suppose the northbound car is traveling at 40 miles per hour, and after two hours, the two cars are 100 miles apart. How fast is the other car going? Explain how you found your answer.

Hours	Distance traveled by northbound car (miles)	Distance traveled by eastbound car (miles)	Distance apart (miles)
1	60	50	$\sqrt{60^2 + 50^2} \approx 78.1$
2	120	100	$\sqrt{120^2 + 100^2} \approx 156.2$
3	180	150	$\sqrt{180^2 + 150^2} \approx 234.3$
4	240	200	$\sqrt{240^2 + 200^2} \approx 312.4$
n	$60n$	$50n$	$78.1n$

6. The bottom of the box has sides of length 6 cm and 7 cm, and $6^2 + 7^2 = 85$, so the diagonal of the bottom of the box has length $\sqrt{85}$ cm. Using this as a leg of the right triangle with hypotenuse d, $d^2 = (\sqrt{85})^2 + (\sqrt{111})^2 = 85 + 111 = 196$, so $d = \sqrt{196} = 14$ cm.

7. See table below left. The distances are changing linearly, and the distance in miles between the cars is 78.1 × the number of hours traveled. Some students may reason as follows: The distance traveled by the northbound car increases by 60 miles every hour; the distance traveled by the eastbound car increases by 50 miles every hour. The distance between them is increasing by 78.1 miles every hour. (Note: Students will probably calculate the distance apart by adding the sum of the squares and taking the square root of that sum.)

8. After 2 hours, the northbound car has traveled 80 miles. Using this distance as one leg of a right triangle and the distance apart (100 miles) as the hypotenuse, and the fact that $100^2 - 80^2 = 3600$, the distance the eastbound car has traveled must be $\sqrt{3600} = 60$ miles. This distance was traveled in 2 hours, so the eastbound car is traveling at 30 mph.

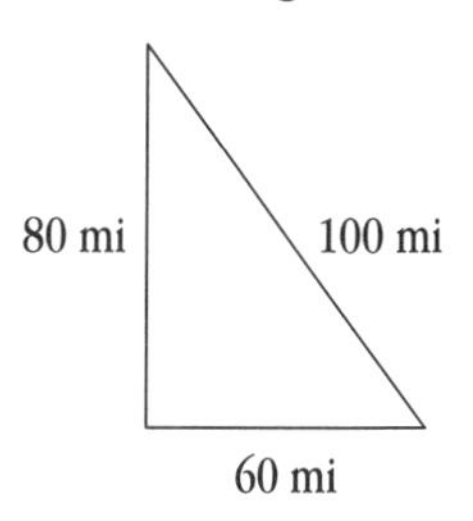

Extensions

9. See below right.

10a. The half-circle *on the leg of length 3* has radius 1.5 and area $\frac{1}{2} \times \pi \times 1.5^2 \approx 3.5$ square units. The half-circle *on the leg of length 4* has radius 2 and area $\frac{1}{2} \times \pi \times 2^2 \approx 6.3$ square units. The half-circle *on the hypotenuse* has radius 2.5 and area $\frac{1}{2} \times \pi \times 2.5^2 \approx 9.8$ square units.

10b. The sum of the areas of the half-circles on the legs is equal to the area of the half-circle on the hypotenuse: $3.5 + 6.3 = 9.8$.

11a. See page 52i.

11b. The sum of the areas of the equilateral triangles on the legs is equal to the area of the equilateral triangle on the hypotenuse: $3.9 + 6.9 = 10.8$.

Extensions

9. Find the perimeter of triangle *ABC.* (Hint: Look for similar triangles.)

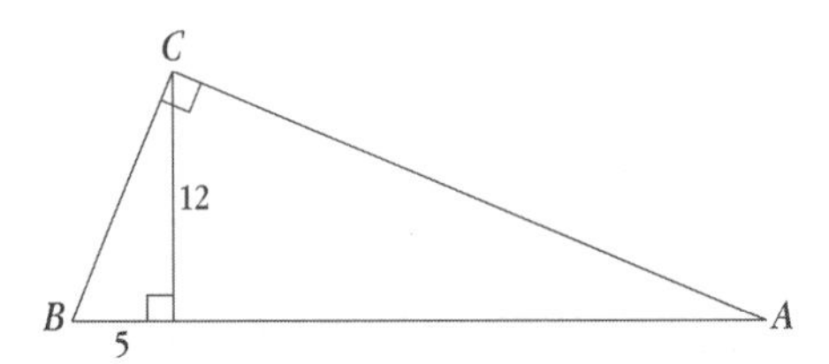

The Pythagorean Theorem describes the relationship among the areas of squares drawn on the sides of a right triangle. In questions 10–12, you will look for relationships among the areas of other shapes drawn on the sides of a right triangle.

10. Half-circles are drawn on the sides of this right triangle.

a. Find the area of each half-circle.

b. How are the areas of the half-circles related?

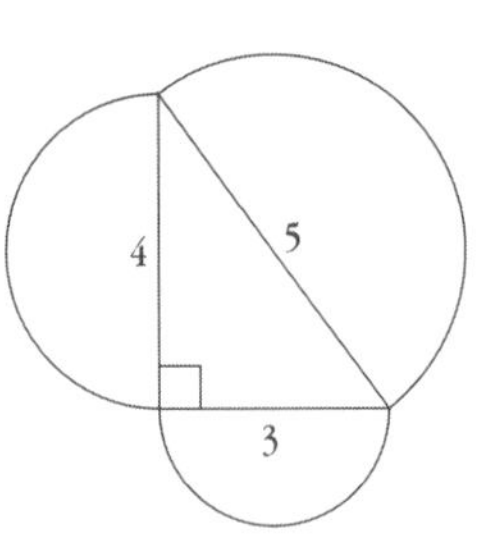

11. Equilateral triangles are drawn on the sides of this right triangle.

a. Find the area of each equilateral triangle.

b. How are the areas of the equilateral triangles related?

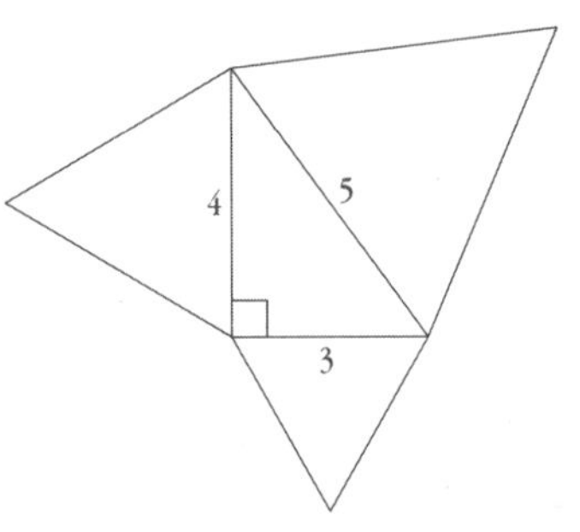

9. The small triangle is similar to triangle *ABC,* as both have angle *B* and a right angle. Since $12^2 + 5^2 = 169$, the length of side *BC* is $\sqrt{169} = 13$. The leg of length 5 on the small triangle corresponds with the leg of length 13 on triangle *ABC,* which means the scale factor from the small to the large triangle is $\frac{13}{5}$, or 2.6. Multiplying the side lengths of the small triangle by 2.6, side *AC* has length $12 \times 2.6 = 31.2$ and side *BA* has length $13 \times 2.6 = 33.8$. The perimeter of triangle *ABC* is thus $13 + 31.2 + 33.8 = 78$. (Note: Students may also calculate that the small triangle has a perimeter of $5 + 12 + 13 = 30$ and that, applying the scale factor, the perimeter of triangle *ABC* is $30 \times 2.6 = 78$.)

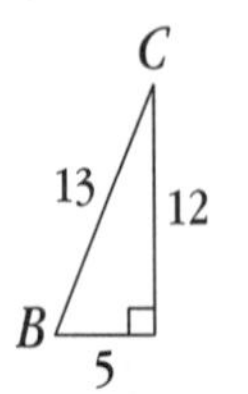

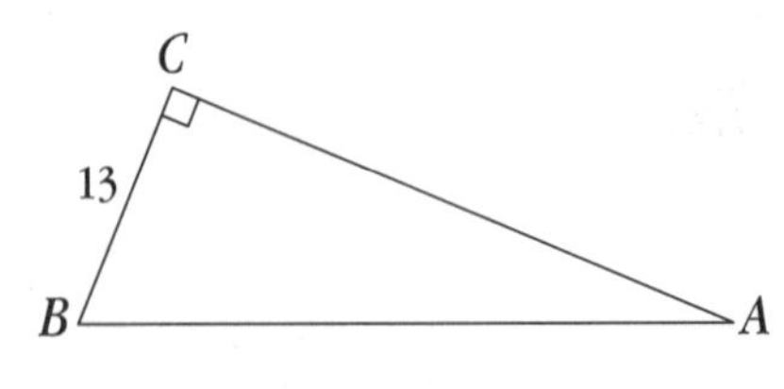

12. Regular hexagons are drawn on the sides of this right triangle.

a. Find the area of each hexagon.

b. How are the areas of the hexagons related?

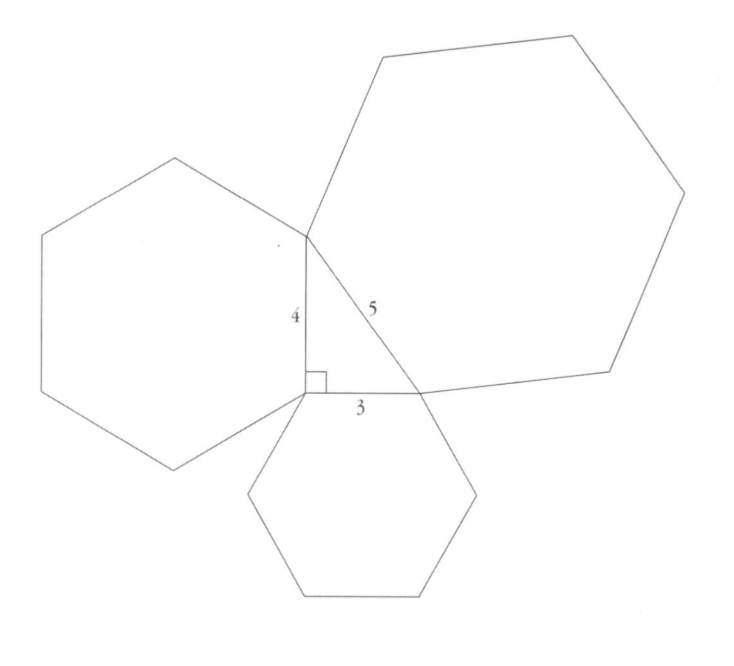

12a. Each hexagon can be divided into six equilateral triangles, the areas of which were found in ACE question 11. The hexagon *on the leg of length 3* has an area of about $6 \times 3.9 =$ 23.4 square units. The hexagon *on the leg of length 4* has an area of about $6 \times 6.9 = 41.4$ square units. The hexagon *on the hypotenuse* has an area of about $6 \times 10.8 =$ 64.8 square units.

12b. The sum of the areas of the hexagons on the legs is equal to the area of the hexagon on the hypotenuse: 23.4 + 41.4 = 64.8.

Mathematical Reflections

In this investigation, you applied the ideas from the first three investigations. The following questions will help you summarize what you have learned:

1. In what ways is the Pythagorean Theorem useful? Give at least two examples.
2. Describe the special properties of a 30-60-90 triangle.

Think about your answers to these questions, discuss your ideas with other students and your teacher, and then write a summary of your findings in your journal.

Possible Answers

1. The Pythagorean Theorem is useful for finding the distance between two points when the coordinates of the points are known. We connect the points with a line segment and then use the segment as the hypotenuse of a right triangle. We draw the two legs, find their lengths, and then find the sum of the squares of the lengths. The distance between the two points is the square root of this sum. The theorem can also be used to find the length of one side of a right triangle if the lengths of the two other sides are known.

2. In a 30-60-90 triangle, the length of the side opposite the 30° angle is half the length of the hypotenuse. The length of the longer leg is $\sqrt{3}$ times the length of the leg opposite the 30° angle. (Note: In a 30-60-90 triangle, if the leg opposite the 30° angle has length a, then the hypotenuse has length 2*a*. So, the longer leg has length $\sqrt{4a^2 - a^2} = \sqrt{3a^2} = a\sqrt{3}$.)

Tips for the Linguistically Diverse Classroom

Diagram Code The Diagram Code technique is described in detail in *Getting to Know Connected Mathematics*. Students use a minimal number of words and drawings, diagrams, or symbols to respond to questions that require writing. Example: Question 2—A student might answer this question by drawing a 30-60-90 triangle and labeling the degrees of each angle. The student could label the side opposite the 30° angle $\frac{1}{2}$ *the hypotenuse* and the longer leg $\sqrt{3}$ *times the length of* with an arrow pointing to the leg opposite the 30° angle.

TEACHING THE INVESTIGATION

4.1 • Stopping Sneaky Sally

In this problem, students apply the Pythagorean Theorem to determine distances on a baseball diamond.

Launch

Introduce the baseball game described in the student edition. Talk about the layout of a baseball diamond, which is pictured on Transparency 4.1. The baseball diamond is a square.

Does anyone know the distance between bases on a standard baseball field? *(90 feet, the length of a side of the square)*

How far do you think a catcher would need to throw the ball to get a runner out at second base?

Let students offer a few estimates, and then have them work in pairs on the problem and follow-up.

Explore

Some students may need help in recognizing the right triangles that are the key to solving the problem and the follow-up.

Suppose you draw a line segment from home plate to second base. What is special about the line segment?

The line segment from home plate to second base is the hypotenuse of a right triangle whose legs are the segments from home plate to first base and from first base to second base.

What do you know about the sides of this right triangle? *(The legs each have a length of 90 feet.)*

How can you find the length of the hypotenuse? *(You can use the Pythagorean Theorem.)*

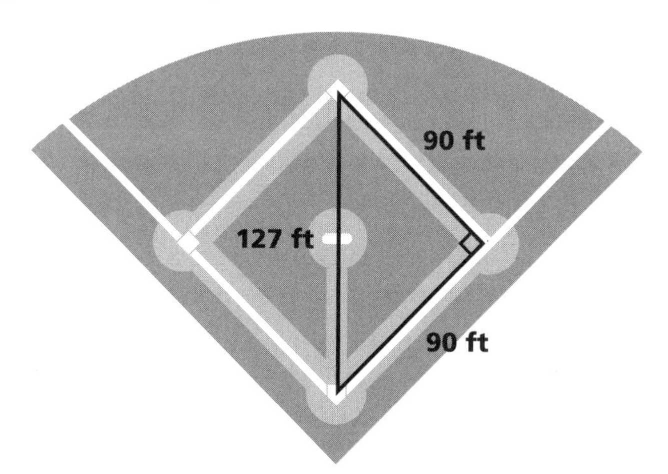

Repeat these questions, if necessary, for the situation in the follow-up. The line segment from home plate to the point halfway between second and third base is the hypotenuse of a right triangle whose legs are the segments from home plate to third base and from third base to the halfway point between second and third base.

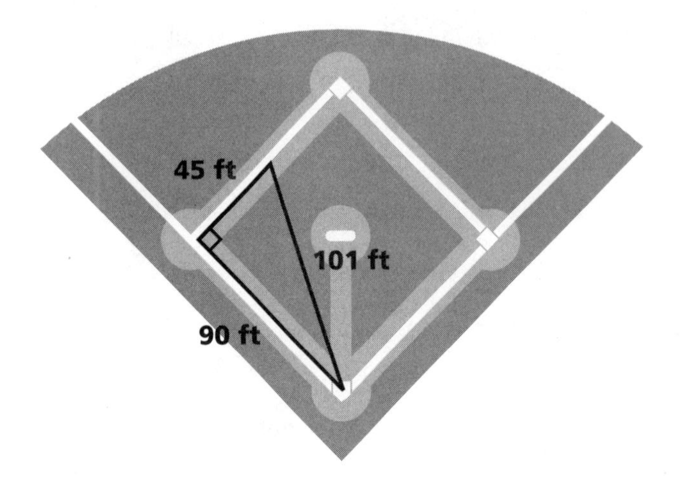

Summarize

Have several students share their strategies for solving the problem and follow-up. Look for specific references to the Pythagorean Theorem.

There are a couple of common misconceptions that may arise during this discussion. First, students may add the lengths of the legs and then square the sum to find the square on the hypotenuse. If this happens, you may need to demonstrate with actual numbers that $(a + b)^2 \neq a^2 + b^2$:

$$(90 + 90)^2 \stackrel{?}{=} 90^2 + 90^2$$

$$180^2 \stackrel{?}{=} 90^2 + 90^2$$

$$32{,}400 \stackrel{?}{=} 8100 + 8100$$

$$32{,}400 \neq 16{,}200$$

A second misconception involves taking square roots: some students will try to find the length of the hypotenuse by calculating $\sqrt{a^2} + \sqrt{b^2}$ rather than $\sqrt{a^2 + b^2}$. Again, offering numerical examples will help them to understand that these expressions are not equivalent. Stress the correct procedure: students should square each digit first, add the squares, and then take the square root of the sum to get the length of the hypotenuse. (At this stage, it is probably best to avoid the symbolism $\sqrt{a^2 + b^2}$.)

4.2 • Analyzing Triangles

In this problem, students investigate the special properties of an equilateral triangle, a 30-60-90 triangle, and an isosceles right triangle.

Launch

Draw an equilateral triangle on the board or overhead.

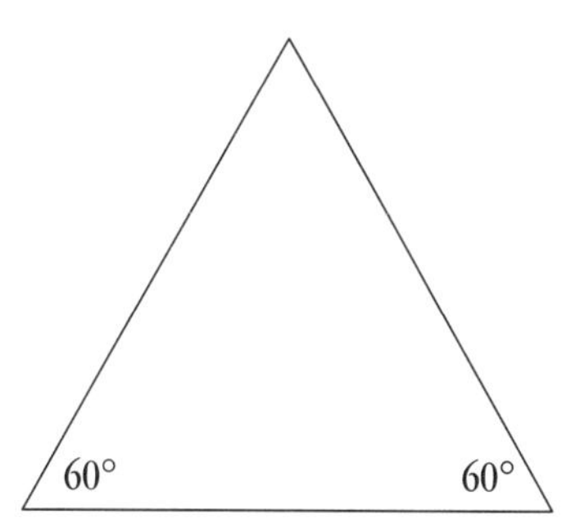

This is an *equilateral triangle.* What is true about the lengths of the sides of an equilateral triangle? *(They are all equal.)*

What is true about the sum of the angles in any triangle? *(It is 180°.)*

What is true about the measures of the angles of an equilateral triangle? *(They are all equal.)*

What is the measure of each angle in an equilateral triangle? *(The sum of the angles in any triangle is 180°, so each angle must measure 60°.)*

Draw the following figure on the board.

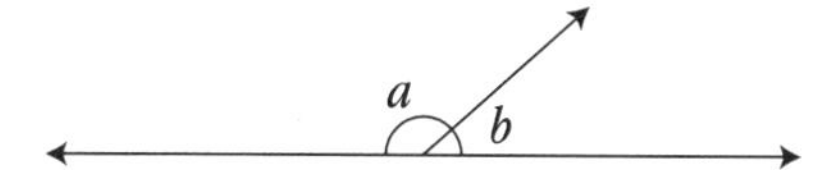

What is the sum of the angle measures in this figure?

Students should recall that the two angle measures add to 180°. (Students discover this fact in the grade 6 unit *Shapes and Designs* and apply it again in the grade 7 unit *Stretching and Shrinking.*) Students will use this idea in solving Problem 4.2.

I want to find the midpoint of the base of the triangle I have drawn. Does anyone have an idea about how I can do this?

One way is to just measure the base to find the halfway point. Another way is to cut a strip of paper the length of the base and fold it in half to locate the midpoint. Mark the midpoint, and draw a vertical line from it to the opposite vertex.

> What does the midpoint do to the base? *(It divides the base into two segments of equal length.)*

Label the angles of the triangle.

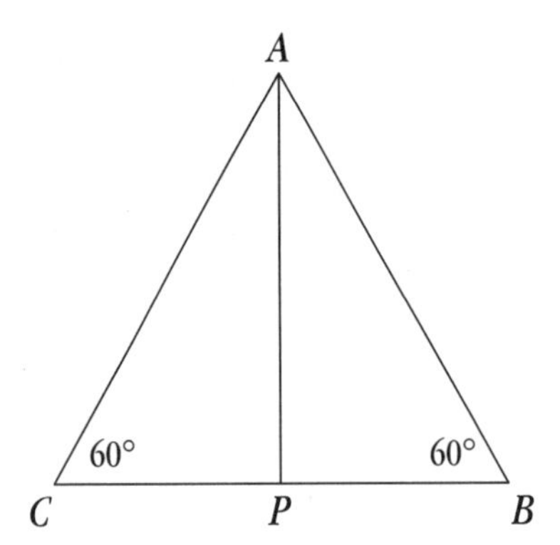

> We have just made two triangles from our equilateral triangle. In Problem 4.2, you will investigate these triangles to discover the relationships between their angles and their side lengths.

Distribute Labsheet 4.2 and scissors, and have students work in pairs on the problem and follow-up.

Explore

Circulate as pairs work. Once students cut out their triangles, have them label the angles and the midpoints by writing the labels on the triangles.

If students are having trouble, ask questions to help them see that two right triangles were formed by cutting the equilateral triangle in half. Then, ask what else they know about these right triangles.

Summarize

Let several pairs share their reasoning about each question, demonstrating their work at the board or overhead. Students should discover that line segment *AP* divides triangle *ABC* into two congruent triangles. You may want to remind students what *congruent triangles* are. Two triangles are congruent if each pair of corresponding sides have the same length. More informally, if one triangle fits on another triangle exactly, or if two triangles have the same size and shape, they are congruent.

Students should be able to reason that both triangles have angles of measure 30°, 60°, and 90°. The midpoint line (also called the *median*) formed two congruent angles along the base of the original equilateral triangle. As the measure sum of angles along a straight line is 180°, the two congruent angles both measure 90°. In each triangle, the larger acute angles measures 60°, so the smaller acute angle measures 30°.

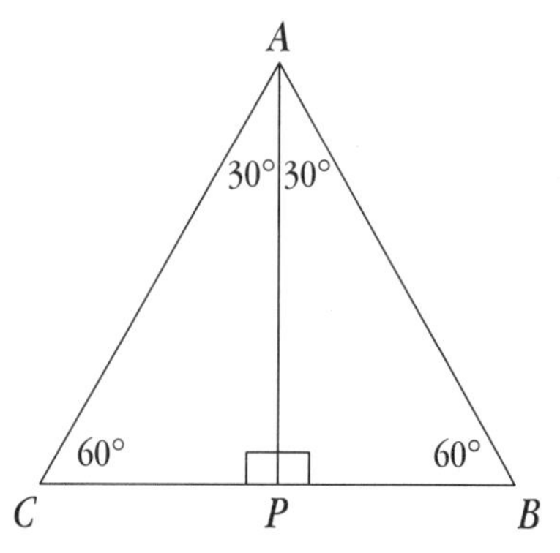

Students should also discover that the length of the side opposite the 30° angle is half the length of the hypotenuse. If not, ask:

> What is the length of segment *CP*? *(Since it is half the length of segment BC, it has a length of 1.)* What is the length of the hypotenuse of right triangle *ACP*? *(2)* The side opposite the 30° angle in this right triangle is half the length of the hypotenuse.

If students have not discovered that the length of the longer leg is $\sqrt{3}$ times the length of the shorter leg, help them to see this. Then ask:

> Suppose you had started with a larger equilateral triangle. Would your discoveries have been different? What if you had started with a smaller equilateral triangle?

In any equilateral triangle, all three angles have the same measure, so the relationships would be the same.

> Would what you discovered be true of *any* 30-60-90 triangle?

You may need to cut out several 30-60-90 triangles to demonstrate that two copies can always be placed back to back to make an equilateral triangle. This is an opportunity to review the properties of similar triangles (which are covered in the grade 7 unit *Stretching and Shrinking*). Students may need help in remembering from their work with concepts of similarity in an earlier grade that all 60-60-60 triangles are equilateral and have the same shape. In similar triangles, the ratios of the lengths of corresponding sides are equal. So, in a 30-60-90 triangle, the ratio of the length of the side opposite the 30° angle to the length of the hypotenuse is always 1 to 2, or $\frac{1}{2}$. If necessary, use other lengths for the sides of the equilateral triangle so students can see that the relationship among the sides remains the same.

Follow-up question 1 reviews the relationships in a 30-60-90 triangle. In question 2, students investigate the relationships in an isosceles right triangle. After going over the question, draw two more isosceles right triangles, and ask students to find the length of the hypotenuse of each triangle. Some students may notice that the hypotenuse is always $\sqrt{2}$ times the length of a side.

For the Teacher: Length Relationships in Special Triangles

Suppose the hypotenuse of a 30-60-90 triangle has length a. The length of the side opposite the 30° angle must be half this length, or $\frac{a}{2}$. The Pythagorean Theorem gives the length of the longer leg as $\frac{a\sqrt{3}}{2}$: the square of the length of the longer leg must be $a^2 - \frac{a^2}{4}$, or $\frac{3a^2}{4}$, so its length is $\sqrt{\frac{3a^2}{4}}$, or $\frac{a\sqrt{3}}{2}$.

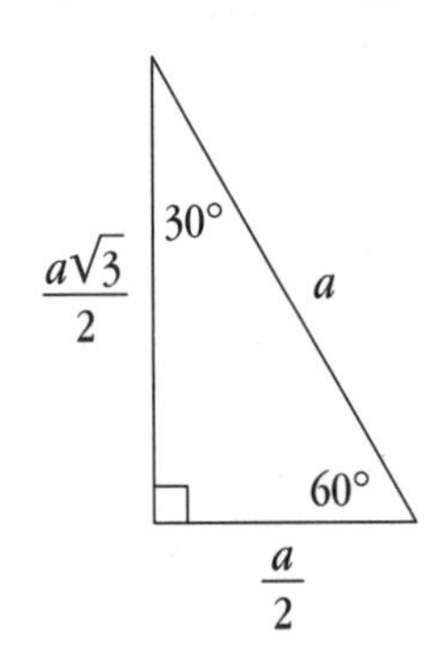

In an isosceles right triangle, the length of the hypotenuse is always the length of one of the legs times $\sqrt{2}$. If the length of each leg is a then, by the Pythagorean Theorem, the square of the length of the hypotenuse must be $a^2 + a^2$, or $2a^2$. Therefore, the length of the hypotenuse is $\sqrt{2a^2} = a\sqrt{2}$.

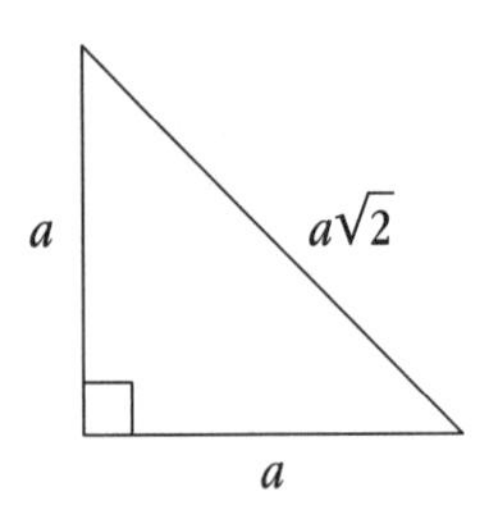

4.3 • Finding the Perimeter

In this problem, students will apply what they have learned about the Pythagorean Theorem and the special properties of 30-60-90 triangles.

Launch

Draw triangle *ABC* on the board, or display Transparency 4.3.

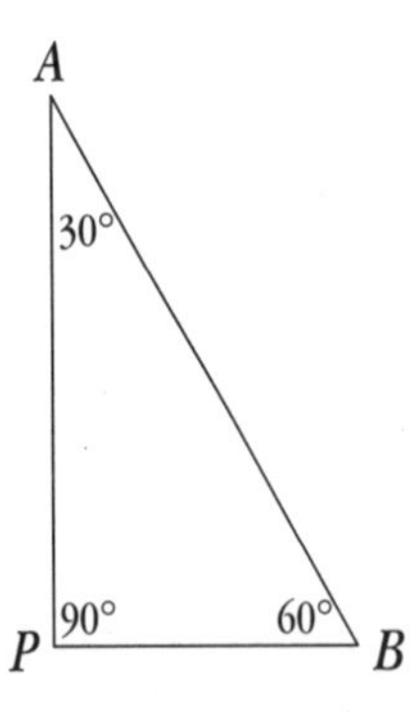

Take a look at triangle *ABC*. What do you need to know to find its perimeter? *(the lengths of the sides)* How can we find those lengths?

Let students offer their ideas. They may notice that the length of the side opposite the 30° angle must be half the length of the hypotenuse but that neither of those two lengths is given. Some may notice that the measure of angle *CAB* is 60°, as the sum of the measures of the other two angles in triangle *ACB* is 120°.

> The challenge for you in this problem is to reason about the relationships in a 30-60-90 triangle and the measures that are given in order to find the side lengths of triangle *ABC* and be able to calculate the perimeter.

Have the class work in groups of four on the problem and follow-up.

Explore

Circulate as groups explore the problem. Some may need help identifying the three 30-60-90 triangles embedded in the figure. Suggest that they redraw the embedded triangles as follows:

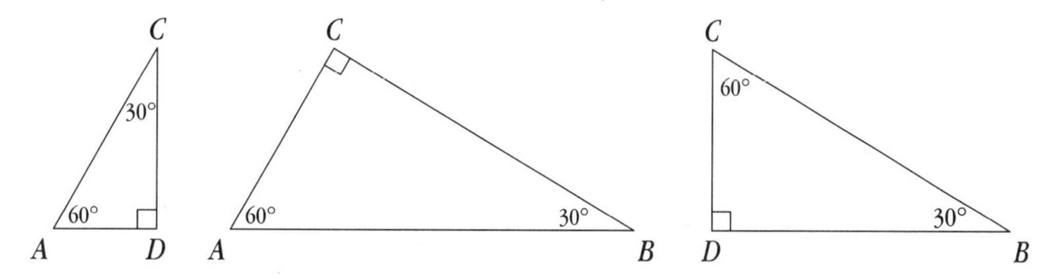

Students may have different strategies for determining the missing measures. Some may start with triangle *BCD,* some with triangle *ABC.*

> How can you find the measure of angle *BCD*? *[This is a right triangle, so the measure is 180° – (90° + 30°) = 60°.]*

> How can you find the measure of angle *CAD*? *(You can use triangle ABC or triangle ACD. In the latter, you will need to find the measure of angle ACD first.)*

Encourage groups to keep track of their calculations in an orderly way so that they will be able to explain their reasoning to the class.

Summarize

Ask one of the groups to describe how they reasoned about the problem. Here is one possible explanation:

- Because the two labeled angles in triangle *ABC* measure 30° and 90°, angle *CAB* must measure 180° – 120°, or 60°. Therefore, angle *ACD* measures 180° – 150° = 30°, and angle *DCB* measures 90° – 30° = 60°.

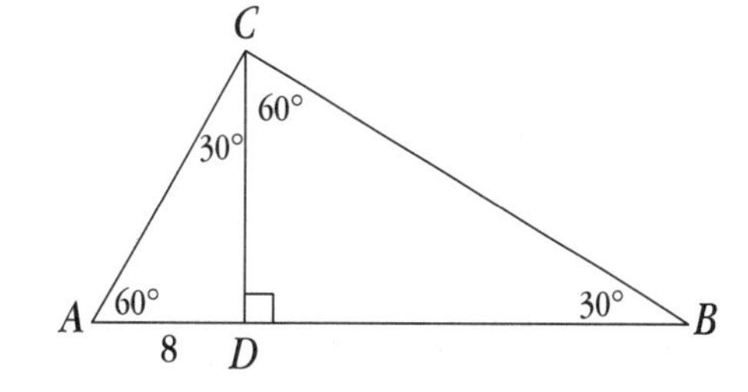

- The side opposite the 30° angle in right triangle *ACD* has a length of 8. The length of the hypotenuse, side *AC*, must be twice that, or 16.

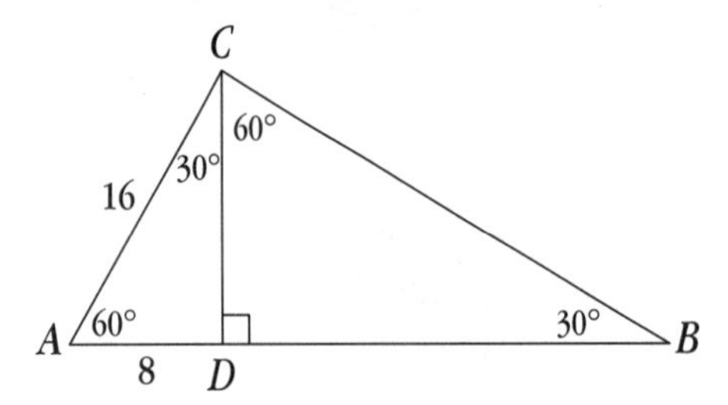

- As side *AB* is the hypotenuse of the 30-60-90 triangle *ABC*, and the length of the side opposite the 30° angle is 16, the length of the hypotenuse, or side *AB*, must be twice that, or 32.

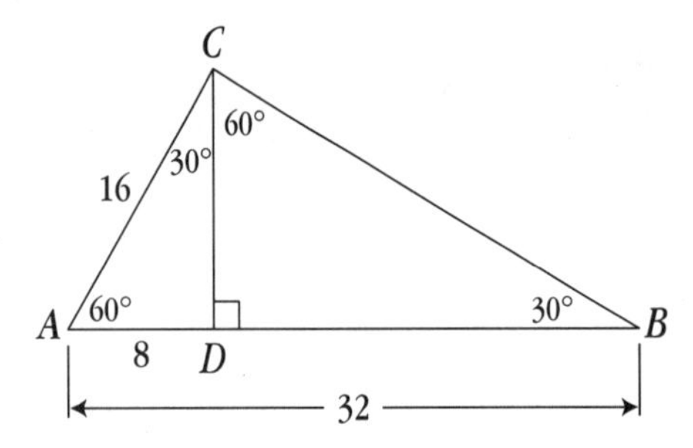

- We can now apply the Pythagorean Theorem to find the missing side length of triangle *ABC*. As one leg and the hypotenuse measure 16 and 32, respectively, the length of side *BC* is the square root of $32^2 - 16^2$, or $\sqrt{768}$.

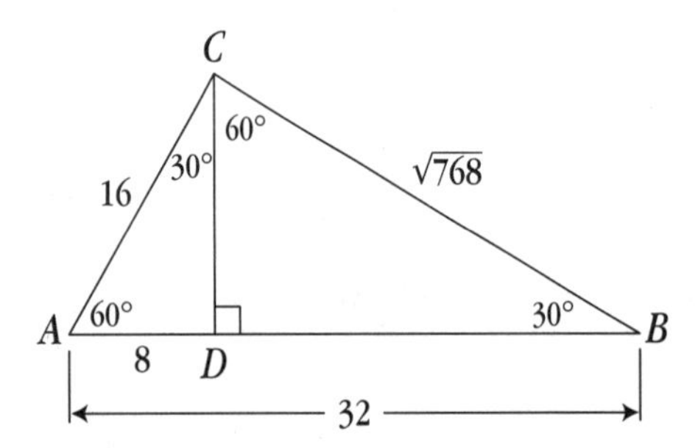

- The perimeter of triangle *ABC* is thus $16 + 32 + \sqrt{768} \approx 16 + 32 + 27.7 \approx 75.7$.

Move on to the follow-up questions. Once students have discussed how they found the areas of the triangles, ask:

> What is the relationship between the areas of the two smaller triangles and the area of the largest triangle? *(The sum of the areas of the two smaller triangles is equal to the area of the largest triangle.)*

Additional Answers

Answers to Problem 4.2

B. Angle *ABP* measures 60° because it is an angle of the original equilateral triangle. Angle *APB* measures 90° because it is half of 180°. The angle measures in any triangle add to 180°, so the remaining angle, angle *PAB,* must measure 30°.

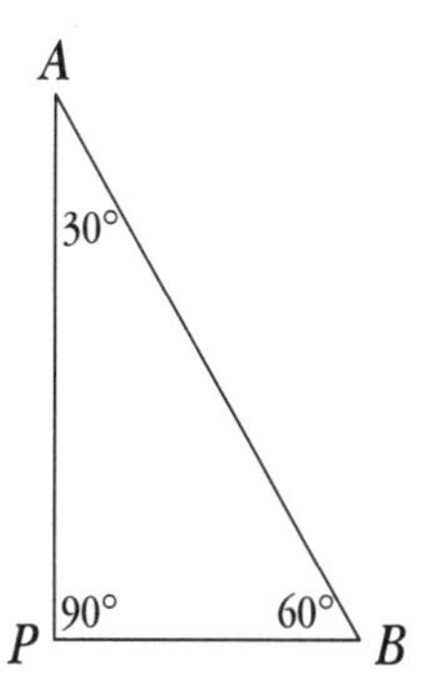

Answers to Problem 4.2 Follow-Up

1. The length of the side opposite the 30° angle is half the length of the hypotenuse, or 3. Since $6^2 - 3^2 = 27$, the length of the other leg is $\sqrt{27}$.

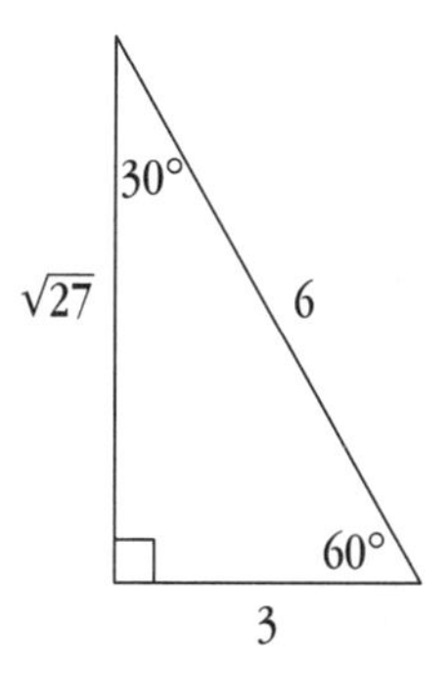

ACE Answers

Extensions

11a. Each equilateral triangle can be divided into two 30-60-90 triangles. The equilateral triangle *on the leg of length 3* is composed of two right triangles, each with a leg of length 1.5 and a hypotenuse of length 3. Since $3^2 - 1.5^2 = 6.75$, the longer leg (which is the height of the equilateral triangle) has length $\sqrt{6.75} \approx 2.6$. This equilateral triangle has an area of about $\frac{1}{2} \times 3 \times 2.6 = 3.9$ square units. The equilateral triangle *on the leg of length 4* is composed of two right triangles, each with a leg of length 2 and a hypotenuse of length 4. Since $4^2 - 2^2 = 12$, the longer leg has length $\sqrt{12} \approx 3.46$. This equilateral triangle has an area of about $\frac{1}{2} \times 4 \times 3.46 = 6.9$ square units. The equilateral triangle *on the hypotenuse* is composed of two right triangles, each with a leg of length 2.5 and a hypotenuse of length 5. Since $5^2 - 2.5^2 = 18.75$, the longer leg has length $\sqrt{18.75} \approx 4.3$. This equilateral triangle has an area of about $\frac{1}{2} \times 5 \times 4.3 = 10.8$ square units.

Irrational Numbers

In this investigation, students take a closer look at square roots. Expressing lengths as decimals leads to a study of decimal representations of fractions and some interesting fraction-decimal patterns.

In Problem 5.1, Analyzing the Wheel of Theodorus, students apply the Pythagorean Theorem to find the lengths of hypotenuses of right triangles. Then, they use a number-line ruler to estimate the lengths. They find, for example, that $\sqrt{5}$ is between 2 and 3 and closer to 2 than 3. Finally, they compare their estimates to those made with a calculator. In Problem 5.2, Representing Fractions as Decimals, students write fractions as terminating or repeating decimals and find fraction equivalents for terminating decimals. In Problem 5.3, Exploring Repeating Decimals, students study the decimal representations of fractions with denominators of 9, 99, and 999, searching for patterns to help them write other decimal equivalents. The problem closes with a short discussion of rational numbers (those that can be represented by either terminating or repeating decimals) and irrational numbers (those that can be represented by nonterminating, nonrepeating decimals).

Mathematical and Problem-Solving Goals

- ***To connect decimal and fractional representations of rational numbers***
- ***To estimate lengths of hypotenuses of right triangles***
- ***To explore patterns in terminating and repeating decimals***

Materials

Problem	For students	For the teacher
All	Graphing calculators	Transparencies: 5.1 to 5.3 (optional)
5.1	Labsheet 5.1 (1 per student), scissors	Transparency of Labsheet 5.1 (optional)

Irrational Numbers

M3RP

In Investigations 2 and 3, you learned methods for finding the lengths of segments connecting dots on a grid. Sometimes you could express the lengths as whole numbers; other times, you had to use $\sqrt{\ }$ symbols or decimal approximations.

The square below has an area of 2 square units. The length of a side of this square can be expressed as $\sqrt{2}$. Just how long is $\sqrt{2}$? If we draw a number line and mark off a segment with the same length as a side of the square, we can see that the length of the segment is about 1.4.

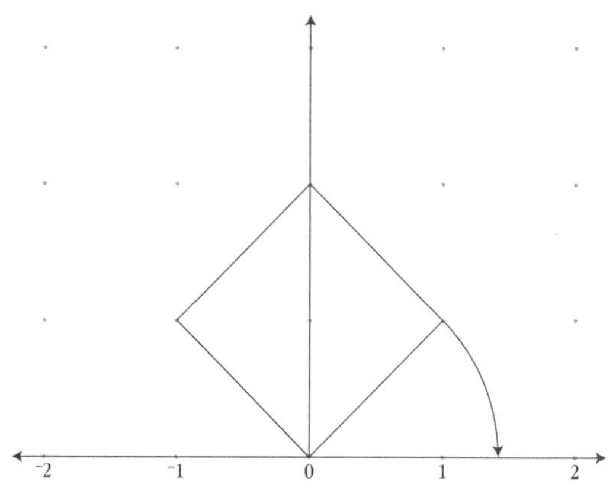

But $\sqrt{2} \times \sqrt{2} = 2$, while $1.4 \times 1.4 = 1.96$. So, 1.4 is too small. If we increase our estimate to 1.42, we get $1.42 \times 1.42 = 2.0164$. So, 1.42 is too large. Try squaring some other decimal numbers to see if you can get a square closer to 2.

If you use your calculator to find $\sqrt{2}$, you get something like 1.414213562, but if you multiply 1.414213562 by 1.414213562 by hand, you get 1.999999998948727844. It seems that even a calculator can't find the exact value of $\sqrt{2}$! So, although you can draw a line segment with a length of $\sqrt{2}$ and locate $\sqrt{2}$ precisely on a number line, it seems extremely difficult to find an exact decimal value for $\sqrt{2}$.

Analyzing the Wheel of Theodorus

At a Glance

Grouping:
small groups

Launch

- Talk with the class about finding a decimal equivalent for $\sqrt{2}$.
- Introduce the Wheel of Theodorus.
- Have groups of two to four work on the problem and follow-up.

Explore

- Have each student label a number-line ruler.
- As students work, check on their understanding of measuring lengths and writing decimals.

Summarize

- Display the Wheel of Theodorus, and ask students for the hypotenuse lengths.
- Talk about decimal equivalents for various square roots.

M³RP

5.1 Analyzing the Wheel of Theodorus

In earlier investigations, you drew segments with lengths of $\sqrt{2}$, $\sqrt{3}$, $\sqrt{4}$, and so on by drawing squares with whole-number areas. In this problem, you will investigate an interesting pattern of right triangles called the Wheel of Theodorus. The pattern will suggest another way to draw segments with lengths that are positive square roots of whole numbers.

The Wheel of Theodorus begins with a triangle with legs of length 1 and winds around counterclockwise. Each triangle is drawn using the hypotenuse of the previous triangle as one leg and a segment of length 1 as the other leg. To make the Wheel of Theodorus, you only need to know how to draw right angles and segments of length 1.

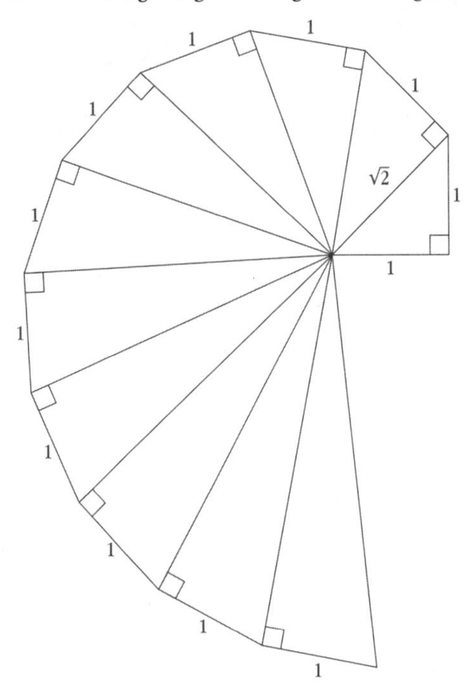

The Wheel of Theodorus is named for its creator, Theodorus of Cyrene. Theodorus was a Pythagorean and one of Plato's teachers.

Assignment Choices

ACE questions 9–20 and unassigned choices from earlier problems

Answers to Problem 5.1

A. The lengths of the hypotenuses from least to greatest are $\sqrt{2}$, $\sqrt{3}$, 2, $\sqrt{5}$, $\sqrt{6}$, $\sqrt{7}$, $\sqrt{8}$, 3, $\sqrt{10}$, $\sqrt{11}$, and $\sqrt{12}$.

B. See page 63g.

C. The lengths $\sqrt{2}$ and $\sqrt{3}$ are between 1 and 2; $\sqrt{5}$, $\sqrt{6}$, $\sqrt{7}$, and $\sqrt{8}$ are between 2 and 3; and $\sqrt{10}$, $\sqrt{11}$, and $\sqrt{12}$ are between 3 and 4.

D. The length $\sqrt{2}$ is between 1.4 and 1.5; $\sqrt{3}$ is between 1.7 and 1.8; $\sqrt{5}$ is between 2.2 and 2.3; $\sqrt{6}$ is between 2.4 and 2.5; $\sqrt{7}$ is between 2.6 and 2.7; $\sqrt{8}$ is between 2.8 and 2.9; $\sqrt{10}$ is between 3.1 and 3.2; $\sqrt{11}$ is between 3.3 and 3.4; and $\sqrt{12}$ is between 3.4 and 3.5.

Problem 5.1

A. Use the Pythagorean Theorem to find the length of each hypotenuse in the Wheel of Theodorus on Labsheet 5.1. Label each hypotenuse with its length. Use the $\sqrt{\ }$ symbol to express lengths that are not whole numbers.

B. Cut out the ruler from Labsheet 5.1. Measure each hypotenuse on the Wheel of Theodorus, and label the point on the ruler that represents its length. For example, the first hypotenuse length would be marked like this:

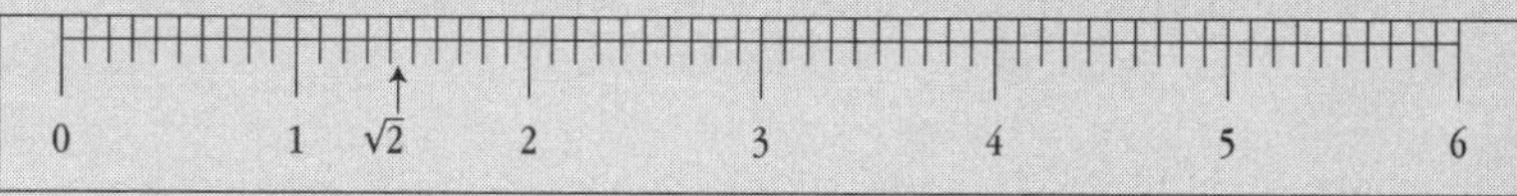

C. For each hypotenuse length that is not a whole number, give the two consecutive whole numbers between which the length is located. For example, $\sqrt{2}$ is between 1 and 2.

D. For each hypotenuse length that is not a whole number, use your completed ruler to find a decimal number that is slightly less than the length and a decimal number that is slightly greater than the length. Try to be accurate to the tenths place.

Problem 5.1 Follow-Up

1. In Problem 5.1, you used a $\sqrt{\ }$ symbol to express the hypotenuse lengths that were not whole numbers. Use your calculator to find the value of each square root, and compare the result to the approximations you found in part D.

2. When Joey used his calculator to find $\sqrt{3}$, he got 1.732050808. Geeta says that Joey's answer must be wrong because when she multiplies 1.732050808 by 1.732050808, she gets 3.000000001. Why do these students disagree?

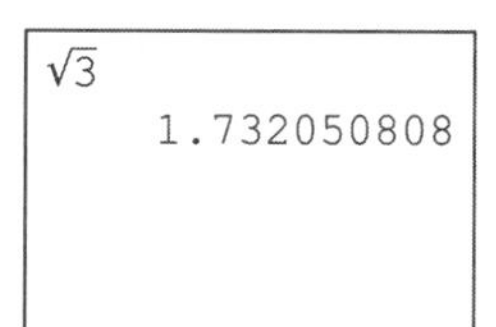

1.732050808*1.73
2050808
3.000000001

Answers to Problem 5.1 Follow-Up

1. $\sqrt{2} = 1.414213562$, $\sqrt{3} = 1.732050808$, $\sqrt{4} = 2$, $\sqrt{5} = 2.236067978$, $\sqrt{6} = 2.449489743$, $\sqrt{7} = 2.645751311$, $\sqrt{8} = 2.828427125$, $\sqrt{10} = 3.16227766$, $\sqrt{11} = 3.31662479$, $\sqrt{12} = 3.464101615$; The numbers obtained using the ruler and the calculator are both approximations, but the calculator gives greater accuracy.

2. Both students have a valid point. Joey's number is an estimate accurate to eight decimal places, while Geeta is correct in pointing out that the square of this decimal approximation does not equal exactly 3. (Note: It would be impossible to write all the decimal places or to indicate any pattern in the decimal expansion for $\sqrt{2}$. This idea will be explored in the next two problems.)

Representing Fractions as Decimals

At a Glance

Grouping: *small groups*

Launch

- Ask for the decimal equivalents for several fractions.
- Introduce the concepts of terminating and repeating decimals.
- Have groups of two or three explore the problem and follow-up.

Explore

- Assist students who have trouble deciding whether a decimal terminates or repeats.

Summarize

- Ask students to explain how they decided whether each decimal terminates or repeats.
- Offer a few more examples.
- Ask if students could predict, without dividing, whether a decimal terminates or repeats.

5.2 Representing Fractions as Decimals

Is it possible to find the exact value of $\sqrt{2}$? Earlier, you found decimal approximations for $\sqrt{2}$. When you multiply such an approximation by itself, you always get a number slightly greater than 2 or slightly less than 2. Even your calculator can't find the exact value of $\sqrt{2}$.

You will get similar results if you try to find an exact value for $\sqrt{3}$ or $\sqrt{5}$. In fact, the decimal representations of these numbers go on forever, and never show any pattern of repeating digits. To better understand these numbers, it helps to compare them to numbers with which you are more familiar: fractions.

In your earlier math work, you learned that any fraction whose numerator and denominator are integers can be expressed as a decimal. The decimal is found by dividing the numerator by the denominator. For example:

$\frac{1}{2} = 0.5$ $\frac{1}{4} = 0.25$ $\frac{1}{3} = 0.333333333333\ldots$ $\frac{1}{11} = 0.090909090909\ldots$

The fractions $\frac{1}{2}$ and $\frac{1}{4}$ have decimal representations that end, or *terminate.* Decimals like these are called **terminating decimals.** The fractions $\frac{1}{3}$ and $\frac{1}{11}$ have decimal representations with groups of digits that repeat forever. Decimals like these are called **repeating decimals.**

Problem 5.2

Write each fraction as a decimal, and tell whether the decimal is terminating or repeating. If the decimal is repeating, tell which digits repeat.

A. $\frac{2}{5}$ **B.** $\frac{3}{8}$ **C.** $\frac{5}{6}$ **D.** $\frac{35}{10}$ **E.** $\frac{8}{99}$

Problem 5.2 Follow-Up

1. Find three fractions with the same decimal representation as the given fraction.
a. $\frac{2}{5}$ **b.** $\frac{5}{6}$

2. Find a fraction that is equivalent to the given terminating decimal.
a. 0.35 **b.** 2.1456 **c.** 89.050

Assignment Choices

ACE questions 1–4, 21–24, and unassigned choices from earlier problems

Answers to Problem 5.2

A. $\frac{2}{5} = 0.4$; terminating
B. $\frac{3}{8} = 0.375$; terminating
C. $\frac{5}{6} = 0.8333\ldots$; repeating; 3
D. $\frac{35}{10} = 3.5$; terminating
E. $\frac{8}{99} = 0.08080808\ldots$; repeating; 08

Answers to Problem 5.2 Follow-Up

A possible answer is offered for each part.

1. a. $\frac{4}{10}, \frac{6}{15}, \frac{8}{20}$ (All have the decimal representation 0.4.)
   b. $\frac{10}{12}, \frac{15}{18}, \frac{20}{24}$ (All have the decimal representation 0.8333 . . . .)
2. a. $0.35 = \frac{35}{100}$ b. $2.1456 = 2\frac{1456}{10,000}$ or $\frac{21,456}{10,000}$ c. $89.050 = 89\frac{50}{1000}$ or $\frac{89,050}{1000}$

5.3 Exploring Repeating Decimals

In the last problem, you saw that any fraction with a numerator and a denominator that are integers can be expressed as a terminating or repeating decimal. It is also true that any terminating or repeating decimal can be expressed as a fraction with a numerator and a denominator that are integers.

Terminating decimals can be written as fractions with a power of 10—such as 10, 100, 1000, and 10,000—in the denominator. For example,

$$0.75 \text{ can be expressed as } \frac{75}{100}, \text{ or } \frac{3}{4}$$

$$1.414213562 \text{ can be expressed as } \frac{1{,}414{,}213{,}562}{1{,}000{,}000{,}000}$$

In this problem, you will explore an interesting pattern that may give you some clues about how to write repeating decimals as fractions.

Problem 5.3

A. Copy the table below, and write each fraction as a decimal.

Fraction	Decimal
$\frac{1}{9}$	
$\frac{2}{9}$	
$\frac{3}{9}$	
$\frac{4}{9}$	
$\frac{5}{9}$	
$\frac{6}{9}$	
$\frac{7}{9}$	
$\frac{8}{9}$	

B. Describe the pattern you see in your table.

C. Use the pattern to write a decimal representation for each fraction. Use your calculator to check your answers.

1. $\frac{9}{9}$ **2.** $\frac{10}{9}$ **3.** $\frac{15}{9}$

D. What fraction is equivalent to each decimal? (Hint: The number 1.222 . . . can be written as 1 + 0.222 . . . .)

1. 1.2222 . . . **2.** 2.7777 . . .

5.3 Exploring Repeating Decimals

At a Glance

Grouping: *pairs*

Launch

- Review with the class how to write terminating decimals as fractions.
- Introduce the idea of writing a repeating decimal as a fraction.
- Have pairs explore the problem and follow-up.

Explore

- Circulate as pairs work, helping students who need assistance.

Summarize

- Talk about the patterns in the decimals for fractions with 9, 99, and 999 in their denominators.
- Verify that students see that the length of the repeating part is equal to the number of 9s in the denominator.
- Ask questions to check students' understanding of the patterns.

Answers to Problem 5.3

A. See page 63g.

B. Each fraction is equivalent to a repeating decimal. The repeating part is a single digit that is equal to the number in the numerator of the fraction.

C. 1. From the pattern, $\frac{9}{9}$ = 0.9999 . . . ; on a calculator, it is equal to 1. (Note: This means that 0.9999 . . . = 1. You may want to talk about this fact with students. See the discussion in the Summarize section.)

2. $\frac{10}{9}$ = 1.1111 . . .

3. $\frac{15}{9}$ = 1.6666 . . .

D. 1. 1.2222 . . . = 1 + 0.2222 . . . = 1 + $\frac{2}{9}$ = $1\frac{2}{9}$

2. 2.7777 . . . = 2 + 0.7777 . . . = 2 + $\frac{7}{9}$ = $2\frac{7}{9}$

Assignment Choices

ACE questions 5–8, 25–31, and unassigned choices from earlier problems

Problem 5.3 Follow-Up

1. Explore the decimal representations of fractions with a denominator of 99. Try $\frac{1}{99}$, $\frac{2}{99}$, $\frac{3}{99}$, and so on. What patterns do you see?

2. Explore the decimal representations of fractions with a denominator of 999. Try $\frac{1}{999}$, $\frac{2}{999}$, $\frac{3}{999}$, and so on. What patterns do you see?

3. Use the patterns you discovered in the decimal representations of fractions with denominators of 9, 99, and 999 to find fractions with these decimal representations.

   a. 0.3333 . . . **b.** 0.05050505 . . . **c.** 0.45454545 . . .

   d. 0.045045045 . . . **e.** 10.121212 . . . **f.** 3.999 . . .

Did you know?

Numbers that can be represented by terminating or repeating decimals are called **rational numbers** because they can be written as a *ratio* of integers. Numbers that can be represented by nonrepeating decimals are called **irrational numbers** because they cannot be written as a ratio of integers. The set of irrational and rational numbers is called the set of **real numbers.**

The number $\sqrt{2}$ is an irrational number. You had trouble finding an exact decimal representation for $\sqrt{2}$ because such a representation does not exist! Other examples of irrational numbers are $\sqrt{3}$, $\sqrt{5}$, and $\sqrt{11}$. In fact, $\sqrt{n}$ is an irrational number for any whole-number value of n that is not a square number.

An amazing fact about irrational numbers is that there is an infinite number of them between any two fractions.

Answers to Problem 5.3 Follow-Up

1. $\frac{1}{99}$ = 0.0101 . . . , $\frac{2}{99}$ = 0.0202 . . . , $\frac{3}{99}$ = 0.0303 . . . , $\frac{10}{99}$ = 0.1010 . . . , $\frac{11}{99}$ = 0.1111 . . . , $\frac{12}{99}$ = 0.1212 . . . ; A fraction with a denominator of 99 is equal to a repeating decimal. The repeating part has two digits: a 0 followed by the number in the numerator if that number is < 10; the number in the numerator if that number is ≥ 10.

2. $\frac{1}{999}$ = 0.001001 . . . , $\frac{2}{999}$ = 0.002002 . . . , $\frac{3}{999}$ = 0.003003 . . . , $\frac{10}{999}$ = 0.010010 . . . , $\frac{11}{999}$ = 0.011011 . . . , $\frac{12}{999}$ = 0.012012 . . . ; A fraction with a denominator of 999 is equal to a repeating decimal. The repeating part has three digits: two 0s followed by the number in the numerator if it is < 10; one 0 followed by the number in the numerator if it is ≥ 10 and < 100; the number in the numerator if it is ≥ 100.

3. a. 0.3333 . . . = $\frac{1}{3}$ b. 0.05050505 . . . = $\frac{5}{99}$ c. 0.45454545 . . . = $\frac{45}{99}$

   d. 0.045045045 . . . = $\frac{45}{999}$ e. 10.121212 . . . = $10\frac{12}{99}$ f. 3.999 . . . = $3\frac{9}{9}$, or 4

Applications • Connections • Extensions

As you work on these ACE questions, use your calculator whenever you need it.

Applications

1. a. Find decimal representations of $\frac{1}{11}$, $\frac{2}{11}$, $\frac{3}{11}$, and so on. Describe the pattern you see.

b. Find three other fractions with the same decimal representations as $\frac{1}{11}$, as $\frac{2}{11}$, and as $\frac{3}{11}$.

In 2–4, write a decimal representation of the fraction.

2. $\frac{30}{9}$ **3.** $\frac{51}{99}$ **4.** $\frac{1000}{999}$

In 5–7, write a fraction representation of the decimal.

5. 0.12121212 . . . **6.** 0. 06060606 . . . **7.** 5.15151515 . . .

8. The decimal below is close to, but between, which two fractions? (Hint: There is more than one answer to this question.)

0.101001000100001 . . .

In 9–12, approximate each square root to four decimal places.

9. $\sqrt{144}$ **10.** $\sqrt{0.36}$

11. $\sqrt{15}$ **12.** $\sqrt{1000}$

In 13 and 14, find the two consecutive whole numbers between which the square root is located. Explain how you found your answer.

13. $\sqrt{27}$ **14.** $\sqrt{1000}$

In 15–17, tell whether the statement is true.

15. $6 = \sqrt{36}$ **16.** $1.5 = \sqrt{2.25}$ **17.** $11 = \sqrt{101}$

14. This is between 31 and 32 because $31 \times 31 = 961$, $32 \times 32 = 1024$, and $\sqrt{1000} \times \sqrt{1000} = 1000$ (which is between 961 and 1024). (Note: Without actually finding the square root of the number, students can use their calculators to square different integers to find two between which the number is located.)

Answers

Applications

1a. $\frac{1}{11} = 0.0909\ldots$, $\frac{2}{11} = 0.1818\ldots$, $\frac{3}{11} = 0.2727\ldots$ $\frac{10}{11} = 0.9090\ldots$; The repeating part consists of two digits and is equal to the numerator of the fraction multiplied by 9.

1b. $\frac{1}{11} = \frac{2}{22} = \frac{3}{33} = \frac{4}{44}$; $\frac{2}{11} = \frac{4}{22} = \frac{6}{33} = \frac{8}{44}$; $\frac{3}{11} = \frac{6}{22} = \frac{9}{33} = \frac{12}{44}$ (**Teaching Tip:** This is a good opportunity to review with students the idea that all equivalent fractions have the same decimal representation.)

2. $\frac{30}{9} = 3.333\ldots$

3. $\frac{51}{99} = 0.5151\ldots$

4. $\frac{1000}{999} = 1.001001\ldots$

5. $0.12121212\ldots = \frac{12}{99} = \frac{4}{33}$

6. $0.06060606\ldots = \frac{6}{99} = \frac{2}{33}$

7. $5.15151515\ldots = 5\frac{15}{99} = 5\frac{5}{33}$

8. Possible answers (there is an infinite number): The decimal is between $\frac{1}{10}$ and $\frac{2}{10}$, between $\frac{10}{100}$ and $\frac{11}{100}$, and between $\frac{101}{1000}$ and $\frac{102}{1000}$.

9. $\sqrt{144} = 12.0000$

10. $\sqrt{0.36} = 0.6000$

11. $\sqrt{15} \approx 3.8730$

12. $\sqrt{1000} \approx 31.6228$

13. This is between 5 and 6 because $5 \times 5 = 25$, $6 \times 6 = 36$, and $\sqrt{27} \times \sqrt{27} = 27$ (which is between 25 and 36).

14. See left.

15. true; $6 \times 6 = 36$

16. true; $1.5 \times 1.5 = 2.25$

17. false; $11 \times 11 = 121$

Connections

18. $\sqrt{11} \approx 3.32$

19. $\sqrt{30} \approx 5.48$

20. $\sqrt{172} \approx 13.11$

21. $12^2 - (\pi \times 6^2) \approx 144 - 113.1 \approx 30.9$ square units

22. 30.9 square units; The four quarter-circles can be combined to make a circle with radius 6 and the problem becomes the same as that in question 21.

Connections

Think about this!

You can estimate the square root of a number without using the square root key on your calculator. Here's how one student thought about estimating $\sqrt{5}$:

- First, I need to find the whole numbers that $\sqrt{5}$ is between. Since $2 \times 2 = 4$ and $3 \times 3 = 9$, I know that $\sqrt{5}$ is between 2 and 3, but it's much closer to 2.
- Next, I need to look at numbers with one decimal place. When I square 2.1, I get 4.41, which is too small. When I square 2.2, I get 4.84, which is closer to 5 but still too small. Squaring 2.3 gives me 5.29, which is too big. Now I know that $\sqrt{5}$ is between 2.2 and 2.3, but it's closer to 2.2. I could use 2.2 as my estimate, but I want to get even closer.
- To get a better estimate, I want to try numbers with two decimal places: $2.21 \times 2.21 = 4.8841$; $2.22 \times 2.22 = 4.9284$; $2.23 \times 2.23 = 4.9729$; $2.24 \times 2.24 = 5.0176$. Now I know that $\sqrt{5}$ is between 2.23 and 2.24. Since 5.0176 is closer to 5 than 4.9284 is, I will use 2.24 as my estimate.

In 18–20, use a method similar to that described in the "Think about this!" box above to estimate the square root to the nearest hundredth. Show your work.

18. $\sqrt{11}$ **19.** $\sqrt{30}$ **20.** $\sqrt{172}$

In 21–24, find the area of the shaded region.

21.

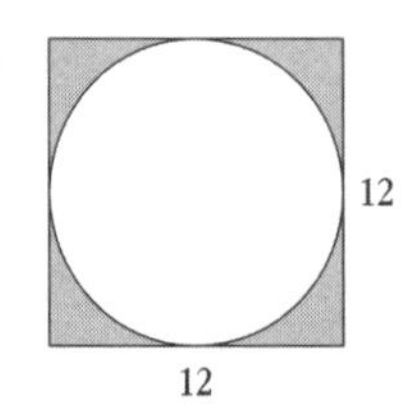

22.

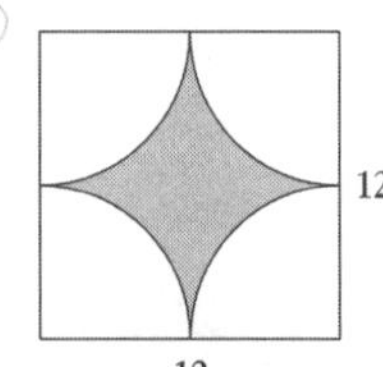

$12^2 - 6^2\pi$
$= 30.9$ u

$13^2 = 169$
$14^2 = 196$

23. 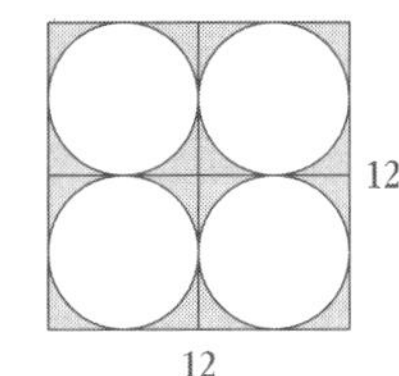

24. 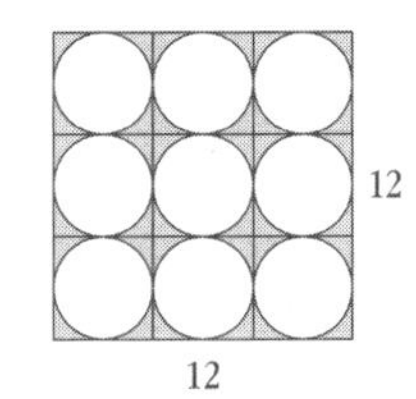

Extensions

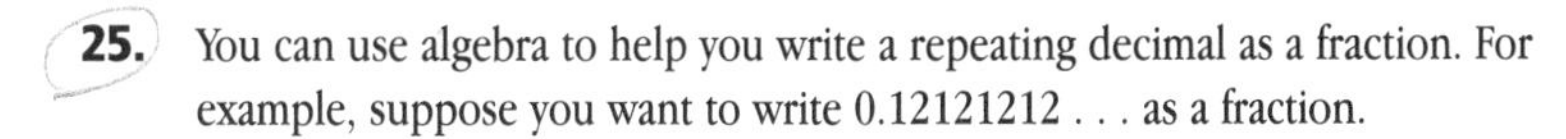

25. You can use algebra to help you write a repeating decimal as a fraction. For example, suppose you want to write 0.12121212 . . . as a fraction.

Let $x = 0.12121212 \ldots$. Multiply both sides of this equation by 100 to get $100x = 12.12121212 \ldots$. Then, subtract the first equation from the second:

$$\begin{array}{rl} 100x = & 12.12121212\ldots \\ - \quad x = & 0.12121212\ldots \\ \hline 99x = & 12 \end{array}$$

Divide both sides of the resulting equation, $99x = 12$, by 99. This gives $x = \frac{12}{99}$. So, $0.12121212\ldots = \frac{12}{99}$.

The key to this method is to multiply each side of the original equation by a power of 10, such as 10, 100, or 1000. You need to use the power of 10 that will shift the decimal point so that one group of the repeating digits is to the left of the decimal point. In the example above, 100 was chosen because the repeating part consisted of two decimal places. Multiplying by 100 shifted the decimal point two places to the right.

Use this method to write each decimal as a fraction.

a. 0. 15151515 . . . **b.** 0. 7777 . . . **c.** 0.123123123123 . . .

23. $4 \times [6^2 - (\pi \times 3^2)] \approx 4 \times [36 - 28.27] \approx 30.9$ square units

24. $9 \times [4^2 - (\pi \times 2^2)] \approx 9 \times [16 - 12.57] \approx 30.9$ square units **(Teaching Tip:** You may want to explore with the class why the shaded area of the four figures in questions 21–24 is the same.)

Extensions

25. See below left.

25a.

$$\begin{array}{rl} 100x = & 15.15151515\ldots \\ - \quad x = & 0.15151515\ldots \\ \hline 99x = & 15 \end{array} \qquad x = \frac{15}{99} = \frac{5}{33}$$

25b.

$$\begin{array}{rl} 10x = & 7.7777\ldots \\ - \quad x = & 0.7777\ldots \\ \hline 9x = & 7 \end{array} \qquad x = \frac{7}{9}$$

25c.

$$\begin{array}{rl} 1000x = & 123.123123123123\ldots \\ - \quad x = & 0.123123123123\ldots \\ \hline 999x = & 123 \end{array} \qquad x = \frac{123}{999} = \frac{41}{333}$$

Note: To check the numbers given in 26–29, students will need to square all three and check whether the two smaller numbers sum to the greatest, as the sides of a right triangle must satisfy the relationship $a^2 + b^2 = c^2$.

26. Since $9^2 + 12^2 = 15^2$, this is a right triangle.

27. Since $5^2 + 7^2 = (\sqrt{74})^2$, this is a right triangle.

28. Since $(\sqrt{2})^2 + (\sqrt{7})^2 = 3^2$, this is a right triangle.

29. Since $6^2 + 10^2 \neq 12^2$, this is not a right triangle.

30. Since $5^2 + 4^2 = 25 + 16 = 41$, but $(5 + 4)^2 = 9^2 = 81$, the expressions are not equal. (**Teaching Tip:** Another way for students to understand that these are not equivalent is to attach this example to the lengths of a right triangle. If these expressions were equal, that would indicate that there exists a right triangle with legs of length 5 and 4 and a hypotenuse of length 9. In a triangle, the length of any one side must be less than the sum of the other two. This is not the case with this set of side lengths.)

31. Since $(\frac{1}{7})^2 = \frac{1}{49}$ and $(\frac{1}{6})^2 = \frac{1}{36}$, the square root of any number between $\frac{1}{49}$ and $\frac{1}{36}$ will be an irrational number between $\frac{1}{7}$ and $\frac{1}{6}$—for example, $\sqrt{\frac{1}{43}}$ or $\sqrt{0.025}$.

In 26–29, tell whether the triangle with the given side lengths is a right triangle, and explain how you know.

26. 9, 12, 15

27. 5, 7, $\sqrt{74}$

28. $\sqrt{2}$, $\sqrt{7}$, 3

29. 6, 10, 12

30. Does $5^2 + 4^2 = (5 + 4)^2$? Explain.

31. Find an irrational number between $\frac{1}{7}$ and $\frac{1}{6}$.

$(\frac{1}{7})^2 = \frac{1}{49}$

$\sqrt{\frac{1}{48}}$ $\sqrt{\frac{1}{47}}$ . . . $\sqrt{\frac{1}{37}}$

$(\frac{1}{6}) = \frac{1}{36}$

.142857

.151551555l...

.166....

Mathematical Reflections

In this investigation, you looked at decimal representations of fractions and fraction representations of decimals. You discovered that some decimals do not terminate or repeat and cannot be represented by fractions with numerators and denominators that are integers. These questions will help you summarize what you have learned:

1. How can you determine whether a given decimal can be written as a fraction?
2. Give three examples of fractions whose decimal representations terminate.
3. Give three examples of fractions whose decimal representations repeat.
4. Give three examples of irrational numbers greater than 5.
5. The positive square root of 10 is an irrational number.

   a. On dot paper, draw a square with sides of length $\sqrt{10}$. What is the area of the square?

   b. Find a fraction close to but less than $\sqrt{10}$. How can you tell that your fraction is less than $\sqrt{10}$?

   c. Find a fraction close to but greater than $\sqrt{10}$. How can you tell that your fraction is greater than $\sqrt{10}$?

Think about your answers to these questions, discuss your ideas with other students and your teacher, and then write a summary of your findings in your journal.

Tips for the Linguistically Diverse Classroom

Original Rebus The Original Rebus technique is described in detail in *Getting to Know Connected Mathematics.* Students make a copy of the text before it is discussed. During the discussion, they generate their own rebuses for words they do not understand; the words are made comprehensible through pictures, objects, or demonstrations. Example: Question 3—Key words and phrases for which students might make rebuses are *three* (3), *fractions* ($\frac{a}{b}$), *decimal representations* ($\frac{1}{2}$ = 0.5), *repeat* (0.33333).

Possible Answers

1. If a decimal terminates or repeats, it can be written as a fraction. (Note: The fraction will have a denominator of 10, 100, 1000, and so on, if the decimal terminates and a denominator of 9, 99, 999, and so on, if the decimal repeats. For example, 0.456 terminates and can be written as $\frac{456}{1000}$; 0.133133 . . . repeats and can be written as $\frac{133}{999}$.)

2. $\frac{1}{2}$, $\frac{1}{4}$, and $\frac{1}{5}$ (Note: Any fraction whose denominator has only the factors of 2 and/or 5 will terminate.)

3. $\frac{1}{3}$, $\frac{1}{9}$, and $\frac{1}{11}$

4. $\sqrt{29}$, $\sqrt{30}$, and 5.1234567891011 . . .

5a. The area is 10 square units.

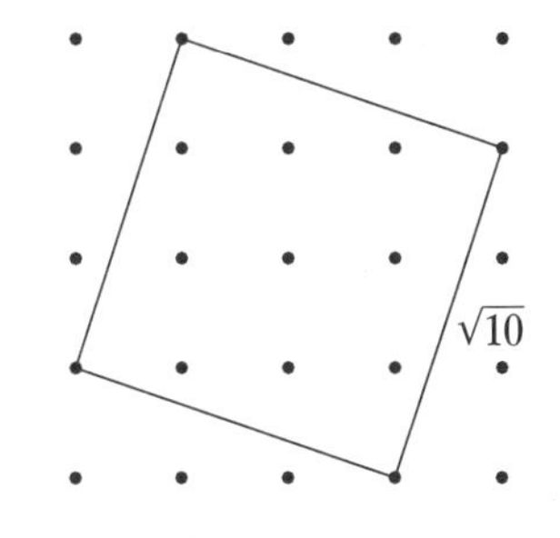

5b. The fraction $\frac{31}{10}$ is less than $\sqrt{10}$. If you square it, you get $\frac{961}{100}$, or 9.61, which is less than 10.

5c. The fraction $\frac{32}{10}$ is greater than $\sqrt{10}$. If you square it, you get $\frac{1024}{100}$, or 10.24, which is greater than 10.

TEACHING THE INVESTIGATION

5.1 • Analyzing the Wheel of Theodorus

In this problem, students explore a mathematically intriguing pattern of triangles, the Wheel of Theodorus. In this series of right triangles, the longer leg of each triangle is the hypotenuse of the preceding triangle. Students apply the Pythagorean Theorem to find the length of the hypotenuse of each triangle in the wheel. Then, they estimate the hypotenuse lengths with a number-line ruler and compare the estimates with those made with a calculator.

Launch

Introduce the topic by discussing how to find a decimal approximation for a square root.

> Think back to when we found the lengths of sides of tilted squares. We first tried to measure the side of a square that had an area of 2 square units. The length had to be a number we could multiply by itself to get 2. Later, we used the $\sqrt{\ }$ key on the calculator to help us find this length.
>
> Just how large is $\sqrt{2}$? Can you find a decimal number that is equal to $\sqrt{2}$? Where is $\sqrt{2}$ on the number line?

Draw a simple number line on the board, and ask students where $\sqrt{2}$ should be placed.

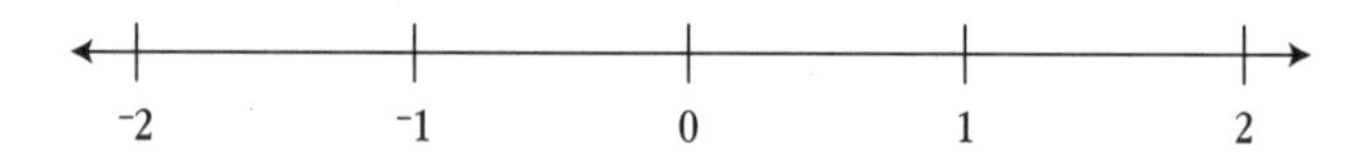

To answer this, students will have to consider about where 1.4 and 1.5 are on the number line. Encourage them to briefly discuss the difficulty of placing a number on a number line when the number cannot be written as an exact decimal: you know what decimals it is between, but not exactly where it is in that interval. For example, is it closer to 1.4 or 1.5?

Draw a square with area 2 on a number line as shown in the student edition.

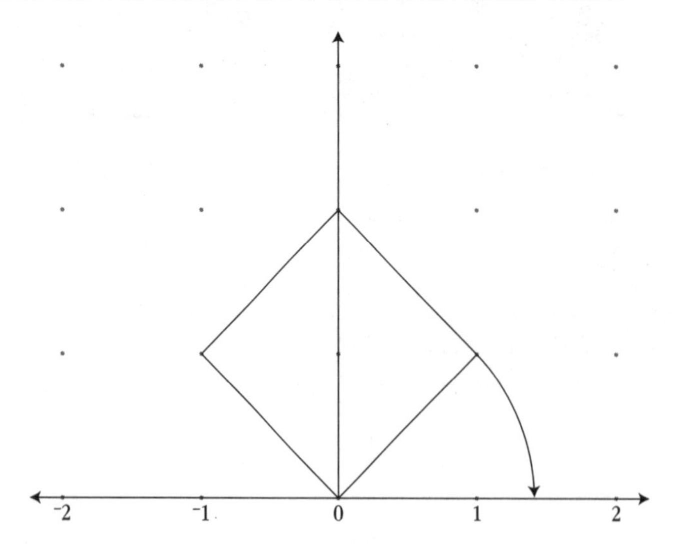

What is the length of a side of this square? *($\sqrt{2}$)*

If we mark off a segment on the number line with the same length as the length of a side, where will the segment end? *(at about 1.4)*

So, $\sqrt{2}$ is approximately equal to 1.4. Does 1.4 = $\sqrt{2}$? *(No, because $1.4^2 = 1.96$.)*

Suppose we try 1.41. Does 1.41 = $\sqrt{2}$? *(No, it is too small; $1.41^2 = 1.9881$.)*

Try 1.42. Does it equal $\sqrt{2}$? *(No, it is too large: $1.42^2 = 2.0164$.)*

Can you find a number that is closer to $\sqrt{2}$ than 1.41 and 1.42 are?

Students should try numbers between 1.41 and 1.42, such as 1.415, 1.413, and 1.414.

Display the Wheel of Theodorus, perhaps projecting a transparency of Labsheet 5.1. Explore with the class how the wheel was constructed, and ask for the lengths of the second and third hypotenuses. Cut out the number-line ruler, and demonstrate how to transfer these lengths to the ruler.

Distribute Labsheet 5.1 and scissors to each student, and have students work in groups of two to four on the problem and follow-up.

Explore

Ask that each student label his or her own number-line ruler. As students work, take the opportunity to check on their understanding of measuring lengths and writing decimals.

Summarize

Display the Wheel of Theodorus. Ask for the lengths of the hypotenuses, and write them on the wheel. Then, have students come to the front and mark the length of each hypotenuse on the number-line ruler.

Ask for approximations to the nearest tenth for each length. As a class, check each approximation by squaring it on a calculator.

Is this estimate too great? Too small? What might be a better estimate? How do you know?

Students should square any additional decimal estimates and compare them to the square of the length of the hypotenuse. Take this opportunity to assess students' understanding of the ordering of decimals. Students sometimes need to review and practice comparing such numbers as 1.41, 1.415, and 1.42.

Ask students to compare their estimates to the numbers they obtained with a calculator in the follow-up. Calculators display varying numbers of decimal places, but students are usually convinced that, no matter what decimal number their calculators display for $\sqrt{2}$, they have not found an exact decimal equivalent. (Note: On many calculators, if the approximation is not cleared before it is squared, the calculator will display the original square as the answer.)

5.2 • Representing Fractions as Decimals

In this problem, students review writing fractions as decimals and decimals as fractions. Then they are introduced to the concepts of terminating decimals and repeating decimals.

Launch

In the student edition, Problem 5.2 opens by returning to the discussion of finding a decimal approximation for $\sqrt{2}$. (The decimal approximation of $\sqrt{2}$ neither repeats nor terminates, so it is not possible to write it as a rational number. Students are not expected to understand why $\sqrt{2}$ is not rational but to recall how hard it was to find a decimal for $\sqrt{2}$.)

Write the fraction $\frac{1}{4}$ on the board.

> Can you find a decimal that is equivalent to this fraction? *(0.25)* How did you find the decimal?

Many students will know this decimal equivalent without performing the division.

> What decimal is equivalent to $\frac{1}{8}$? *(0.125)* How did you find that? *(by dividing 8 into 1)*
>
> Find a decimal equivalent to $\frac{1}{3}$. *(0.33333 . . .)* What is different about this decimal?

If students have used a calculator to find this decimal, they may think that it terminates. If so, write the decimal 0.333333 as a fraction, $\frac{333,333}{1,000,000}$, and ask the class to compare it to $\frac{1}{3}$. It is worth taking the time to divide 1 by 3 on the board to show that the 3 in the decimal repeats forever, or infinitely.

> Write the fraction $\frac{1}{11}$ as a decimal by dividing 1 by 11 without using your calculator. *(0.09090909 . . .)* What do you notice? *(The 09 repeats as the 3 did for $\frac{1}{3}$.)*
>
> Now, use your calculator to write $\frac{1}{11}$ as a decimal. What do you see?

If students' calculators round the last decimal place, showing 0.0909091 or 0.090909091, talk with the class about why this is happening. Explain that a better indication of the decimal would be 0.09090909 . . . or 0.0909090909 . . . , as this notation shows that the decimal continues forever.

Discuss the information on terminating decimals and repeating decimals presented in the student edition. Challenge students to look for patterns while they work on Problem 5.2 that will help them to predict which fractions have representations that are terminating decimals and which have representations that are repeating decimals.

Have students explore the problem and follow-up in groups of two or three.

Explore

Students can use calculators to find the decimal representation of each fraction. However, they may need assistance to decide which decimals terminate and which repeat, as all decimals may seem to terminate on a calculator. If so, ask:

Do you see the repeating part of the decimal in the calculator display?

Suggest that students convert at least one of the fractions without using a calculator to verify that the decimal does indeed repeat.

The follow-up questions review equivalent fractions and decimals.

Summarize

Ask students for the decimal representation for each fraction and to explain how they decided whether the decimal terminates or repeats. Evaluate one or two of the fractions without using a calculator to show the repeating part more clearly. Then, offer the class a similar problem:

Without using your calculator, write the fraction $\frac{4}{33}$ as a decimal. *(0.12121212 . . .)* Does the decimal repeat or terminate? *(It repeats; the repeating part is 12.)*

The three ellipsis dots indicate that a pattern continues infinitely, as in 0.21342134 . . . .

Have students explore other fractions with their calculators until they realize that no matter how unusual the fraction seems, they will always get a repeating decimal or a terminating decimal.

Without dividing, do you think that you can predict whether a fraction has a repeating or terminating decimal representation?

Some students will enjoy this challenge and may want time to think about it. For a fraction to terminate, it must be possible to express it as an equivalent fraction with a denominator of a power of 10, such as 10, 100, 1000, and 10,000. This will happen if the fraction has only 2s and 5s as factors of its denominator. For example, $\frac{1}{8} = \frac{125}{1000} = \frac{125}{2 \times 2 \times 2 \times 5 \times 5 \times 5}$, or 0.125.

Go over the follow-up questions. The idea of converting terminating decimals to fractions leads into the topic of Problem 5.3.

5.3 • Exploring Repeating Decimals

In this problem, students search for a method for writing repeating decimals as fractions. Then they are introduced to the concepts of rational and irrational numbers.

Launch

Use the terminating decimals explored in Problem 5.2 to help students review how to write a terminating decimal as a fraction. They may need to be reminded how to count decimal places to find the appropriate power of 10 for the denominator.

Then, write a repeating decimal such as 0.333 . . . on the board. Ask:

What is a fraction representation of this decimal?

Some students may recall that this decimal is equivalent to $\frac{1}{3}$. If some suggest $\frac{3}{10}$, $\frac{33}{100}$, or the like, have them check their answers by converting them to decimals. (Students should know that fractions with powers of 10 in the denominator are decimal fractions—that is, they can be written as decimals directly.)

How could you write the decimal 0.454545 . . . as a fraction?

Students may guess $\frac{45}{100}$ but should recognize that $\frac{45}{100}$ is close but not equal to this decimal. If so, ask the class how to write a fraction that is just a bit greater. Some students may suggest a fraction such as $\frac{45}{101}$; if they use their calculator to check this idea, they will discover that this fraction is even less than $\frac{45}{100}$.

Making the denominator greater does what to a fraction? *(It makes the fraction smaller.)* What do you suggest we try now? *(Make the denominator smaller.)*

Tell students that in today's problem they will look for patterns that can help them to write repeating decimals as fractions. Let them work on the problem and follow-up in pairs.

Explore

The pattern in the fractions with only 9s in their denominators are obvious, so students should find making the table relatively straightforward.

In part C, some students may just divide 15 by 9. Help them to see the patterns in fractions with denominators of 9. If the numerator is greater than the denominator, students must first express the fraction as a mixed number. For example, $\frac{15}{9}$ is $1\frac{6}{9}$ and hence is equal to 1.6666 . . . .

Summarize

Discuss the patterns in the decimals students wrote for fractions with denominators consisting only of 9s. Then ask:

What is the decimal representation for $\frac{19}{9}$? *($2\frac{1}{9}$ = 2.111111 . . .)* For $\frac{20}{9}$? *($2\frac{2}{9}$ = 2.222222 . . .)*

What fraction is equivalent to 4.111111 . . . ? *($4\frac{1}{9}$, or $\frac{37}{9}$)*

For $\frac{9}{9}$ in part C, students will get the repeating decimal 0.999999 . . . when they follow the pattern they found in the problem, which means that 0.999999 . . . = 1. If they have difficulty understanding this idea, you may want to explain that as the 9 repeats infinitely, we can get as close to 1 as we like by continuing to add 9s to the decimal representation. If students insist that the repeating decimal would still be too small to equal 1, ask them to write a decimal that is greater than 0.999999. . . and equal to 1. It is not important that they completely understand this, but they may have fun thinking about these ideas.

Students should have discovered these patterns in their work in the problem and follow-up:

- A fraction with a denominator of 9 has a decimal representation with a one-digit repeating pattern, or *repetend,* and the repetend is the number in the numerator. (You may or may not want to introduce the term *repetend* to your class.)
- A fraction with a denominator of 99 has a decimal representation with a two-digit repetend: a 0 followed by the number in the numerator if that number is less than 10; the number in the numerator if that number is equal to or greater than 10.
- A fraction with a denominator of 999 has a decimal representation with a three-digit repetend: two 0s followed by the number in the numerator if that number is less than 10; one 0 followed by the number in the numerator if that number is equal to or greater than 10 but less than 1000; the number in the numerator if that number is equal to or greater than 100.

Ask questions to check students' understanding of these ideas.

> What is the decimal representation of $\frac{50}{99}$? *(0.505050 . . .)*
>
> What is the decimal representation of $\frac{100}{99}$? *($1\frac{1}{99}$, or 1.010101. . .)*
>
> What fraction is equivalent to 0.353535 . . . ? *($\frac{35}{99}$)*
>
> Can you predict the decimal representations for fractions with denominators of 9999, 99,999, and so on?

Assuming the fractions are proper fractions, the decimal representations will have repetends of three digits, four digits, and so on, that are the same as the numerator of the fraction. A fraction with a denominator of 9999 is equivalent to a repeating decimal with a four-digit repetend. If the numerator is less than 10, the repetend is 000 followed by the digit in the numerator. If the numerator is between 10 and 99, inclusive, the repetend is 00 followed by the number in the numerator. If the numerator is between 100 and 999, inclusive, the repetend is 0 followed by the number in the numerator. If the numerator is between 101 and 9998, inclusive, the repetend is the number in the numerator.

The important pattern for students to note is that *the length of the repeating part is equal to the number of 9s in the denominator.* The repeating part is the number in the numerator with the appropriate number of 0s preceding it to attain this length. The patterns seen with denominators of 9, 99, 999, and so on, can be used to write repeating decimals as fractions.

For the Teacher: Converting Repeating Decimals to Fractions

Another method for converting a repeating decimal to a fraction involves solving an equation. Students can explore this method in ACE question 25. To convert 0.454545 . . . to a fraction, for example, call the unknown fraction *N:*

$$N = 0.454545\ldots$$

Multiply both sides of this equation by 100:

$$100N = 45.454545\ldots$$

Subtract the first equation from the second:

$$\begin{array}{rl} 100N = & 45.454545\ldots \\ -\quad N = & 0.454545\ldots \\ \hline 99N = & 45 \end{array}$$

Therefore,

$$N = \frac{45}{99}, \text{ or } \frac{5}{11}$$

Additional Answers

Answers to Problem 5.1

B. Note that $\sqrt{4} = 2$ and $\sqrt{9} = 3$.

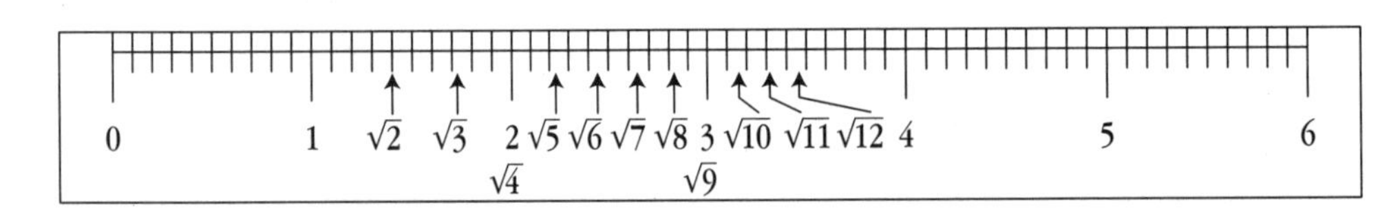

Answers to Problem 5.3

A.

Fraction	Decimal
$\frac{1}{9}$	0.1111 . . .
$\frac{2}{9}$	0.2222 . . .
$\frac{3}{9}$	0.3333 . . .
$\frac{4}{9}$	0.4444 . . .
$\frac{5}{9}$	0.5555 . . .
$\frac{6}{9}$	0.6666 . . .
$\frac{7}{9}$	0.7777 . . .
$\frac{8}{9}$	0.8888 . . .

Rational and Irrational Slopes

In this investigation, students explore an interesting application of irrational numbers. The context is a video game in which the main character, Oskar, is trying to escape from a forest in which the trees are planted in rows.

In Problem 6.1, Revisiting Slopes, students review how to find the slope of a line drawn on a grid and the connection between the slope of a line and points on that line. They relate these concepts to what they know about rational and irrational numbers. Students discover that any ratio of whole-number lengths they choose as a slope for Oskar's path will eventually hit a tree if the forest is large enough; yet, intuitively, they know that *some* straight-line path must clear the trees because the path is infinitely adjustable. The situation is resolved in Problem 6.2, Escaping from the Forest, when students make the connection between slopes that are irrational numbers and the points on a grid. In order for a line not to pass through a tree (or grid dot), the line must have an irrational slope. The relationships between parallel lines (which have the same slope) and perpendicular lines (whose slopes are negative reciprocals) are explored in the ACE questions.

Mathematical and Problem-Solving Goals

- ***To review the concept of the slope of a line***
- ***To connect the concept of slope to the idea of irrational numbers***
- ***To use slopes to test whether lines are parallel or perpendicular***

Materials

Problem	For students	For the teacher
All	Graphing calculators	Transparencies: 6.1A to 6.2 (optional)
6.2	Labsheet 6.2 (1 per student), dot paper	Transparency of dot paper (optional)
ACE	Labsheets 6.ACE1 and 6.ACE2 (optional; 1 each per student), dot paper	

INVESTIGATION 6

Rational and Irrational Slopes

In this unit, you have used the Pythagorean Theorem to find lengths. In the process of finding lengths, you discovered irrational numbers—numbers that cannot be represented as fractions whose numerators and denominators are integers. In this investigation, you will explore an interesting application of irrational numbers.

Caitlin is playing a video game in which she directs a character named Oskar through a series of obstacles. At this point in the game, Oskar is trapped in the center of an immense forest made up of trees planted in rows and columns.

The trees are actually laser beams that are so narrow their widths cannot be measured. The beams are focused straight up. If Oskar hits a laser, he will be destroyed. To help him escape, Caitlin has used most of Oskar's power to reduce him to the width of a laser beam. Now he has only enough power to walk in a straight line. Can Oskar walk in a straight line out of the forest without running into a laser tree?

Tips for the Linguistically Diverse Classroom

Rebus Scenario The Rebus Scenario technique is described in detail in *Getting to Know Connected Mathematics*. This technique involves sketching rebuses on the chalkboard that correspond to key words in the story or information that you present orally. Example: Some key words and phrases for which you may need to draw rebuses while discussing the video game on this page are *video game* (terminal with keyboard), *Oskar* (cartoon character), *forest* (trees), *straight up* (↑), *destroyed* (erase Oskar), *reduce him to the width of a laser beam* (redraw Oskar as a very narrow stick figure), *straight line* (————).

6.1 Revisiting Slopes

To help Oskar escape from the forest, Caitlin must find a straight-line path that does not hit any laser trees. In your earlier work, you used the idea of *slope,* or steepness, to help draw and describe lines. If you imagine a coordinate system with the origin where Oskar is now standing, you can think about the slopes of straight-line paths through the forest. For example, the line OB in the figure below has a slope of $\frac{1}{7}$ because it rises 1 unit for every 7 units it moves to the right.

Problem 6.1

The diagram below shows part of the forest. Some of the laser trees have been labeled with letters, and x- and y-axes have been added. Oskar's location is labeled with an O.

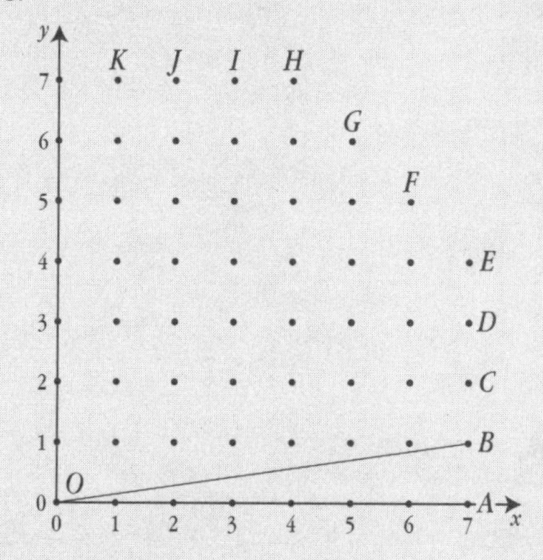

If Oskar had enough power, he could use a laser shield to walk right through the trees. Give the slope of the straight-line path he could follow to get from point O to each of the labeled trees.

Problem 6.1 Follow-Up

1. Suppose the forest continues for a great distance in all directions.
   a. Give the coordinates of two more trees on line OC.
   b. Give the coordinates of two more trees on line OF.

Revisiting Slopes

At a Glance

Grouping: ***pairs***

Launch

- Talk about the video game and Oskar's location in the forest.
- Review finding the slope of a line.
- Have pairs work on the problem and follow-up.

Explore

- Ask questions to verify that students understand the concept of slope.

Summarize

- Have students discuss the slopes they found and how they found them.
- Talk about the follow-up, reviewing the concept of slope and making the connection between points on a grid and rational and irrational numbers.

Answer to Problem 6.1

Line	Slope
OA	0
OB	$\frac{1}{7}$
OC	$\frac{2}{7}$
OD	$\frac{3}{7}$
OE	$\frac{4}{7}$
OF	$\frac{5}{6}$

Line	Slope
OG	$\frac{6}{5}$
OH	$\frac{7}{4}$
OI	$\frac{7}{3}$
OJ	$\frac{7}{2}$
OK	$\frac{7}{1}$

Answers to Problem 6.1 Follow-Up

1. a. Possible answer: (14, 4) and (21, 6)
   b. Possible answer: (12, 10) and (18, 15)

Assignment Choices

ACE questions 1–6, 8, 9, 15, and unassigned choices from earlier problems

Escaping from the Forest

At a Glance

Grouping:
small groups

Launch

- Suggest several rational slopes and ask whether a path along them would hit a tree.
- Talk about the two lines drawn on the grid in the student edition.
- Have groups of three or four explore the problem and follow-up.

Explore

- Help students who need assistance finding a line with slope $\sqrt{2}$.
- Encourage students to explain why a line with an irrational slope won't hit a tree.

Summarize

- Review how to find a line with slope $\sqrt{2}$.
- Discuss the follow-up.

2. **a.** Give the coordinates of three trees on the line with slope 1 that passes through point O.
   b. What angle does the line described in part a make with the x-axis?
3. If a line contains two points on the grid, how can you find its slope?
4. If a line contains two points on the grid, will its slope be a rational number or an irrational number?

6.2 Escaping from the Forest

You have seen that irrational numbers, such as $\sqrt{2}$, $\sqrt{3}$, and $\sqrt{5}$, can represent lengths. Irrational numbers can also represent slopes of lines. Earlier in this unit, you found that $\sqrt{2}$ is between 1.4 and 1.5, or between $\frac{7}{5}$ and $\frac{3}{2}$. This means that the line through point O with slope $\sqrt{2}$ must lie between the line with slope $\frac{7}{5}$ and the line with slope $\frac{3}{2}$.

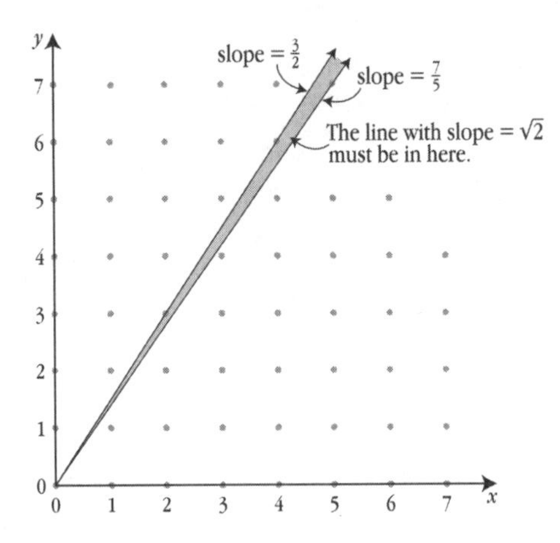

Problem 6.2

The laser forest extends beyond the screen, but Caitlin is not sure how far. She wants to be sure Oskar won't hit a tree anywhere along his escape path.

Study the drawing above. On Labsheet 6.2, draw a straight-line path that Oskar could follow to get out of the forest. Give the slope of the path you find. Explain why you think your path will work.

Assignment Choices

ACE questions 7, 10–14, 16, and unassigned choices from earlier problems

2. a. Possible answer: (1, 1), (2, 2), and (3, 3)
   b. The line makes a 45° angle with the x-axis.
3. The slope is the ratio of the rise to the run, or the ratio of the difference between the two y-coordinates to the difference between the two x-coordinates.
4. The slope of any line that contains two points on the grid is a rational number. It is the ratio of two integers, because the vertical distance (rise) and the horizontal distance (run) are whole numbers.

Problem 6.2 Follow-Up

1. **a.** On dot paper, draw a line with slope $\sqrt{5}$. Explain the steps you use to draw your line.

   b. On the same paper, draw a line with a slope that is a rational number slightly less than $\sqrt{5}$. Draw another line with a slope that is a rational number slightly greater than $\sqrt{5}$. Give the slopes of the lines you draw.

2. In addition to a path with slope $\sqrt{2}$, describe some other paths that Oskar could follow to get out the forest. What is true about the slope of any path he could follow?

Answer to Problem 6.2

Students' explanations should be similar to that given in the Summarize section. That is, find a line with length $\sqrt{2}$ and then use this length to locate the point $(1, \sqrt{2})$. The line from the origin (Oskar) through this point will not hit any trees. [Note: A length of $\sqrt{2}$ could also help locate the point $(\sqrt{2}, 1)$. The line from the origin to the point $(\sqrt{2}, 1)$ will not hit any trees either. The slope of this line is $\frac{1}{\sqrt{2}}$.]

Answers to Problem 6.2 Follow-Up

1. See page 72f.
2. There are many paths that Oskar can follow, such as lines with slopes $\sqrt{3}$, $\sqrt{6}$, and $\sqrt{8}$. All of the paths have a slope that is an irrational number.

Answers

Applications

1a. $\frac{1}{2}$

1b. $\frac{-3}{2}$

1c. $\frac{-1}{1} = -1$

1d. $\frac{6}{2} = 3$

1e. $\frac{4}{2} = 2$

1f. $-\frac{1}{6}$

1g. $-\frac{1}{6}$

1h. $\frac{4}{1} = 4$

1i. $\frac{-3}{1} = -3$

2a. These lines are parallel. Line *AE* has slope $\frac{6}{6} = 1$, and line *JK* has slope $\frac{2}{2} = 1$. The slopes are equal.

2b. These lines are parallel. Line *FK* has slope $-\frac{1}{2}$, and line *HJ* has slope $-\frac{1}{2}$. The slopes are equal.

3. Possible answer: lines *JB* and *KL*, lines *IF* and *LD*; Parallel lines have the same slope because the ratios of rise to run are equal.

Applications • Connections • Extensions

As you work on these ACE questions, use your calculator whenever you need it.

Applications

In 1–6, use this grid. Copy the points onto dot paper if you need to.

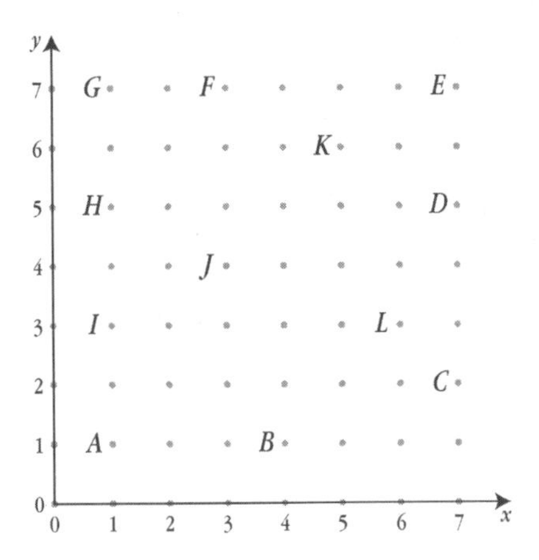

1. Find the slope of each line.

a. line *KE* **b.** line *GJ* **c.** line *LC*

d. line *AF* **e.** line *IF* **f.** line *IC*

g. line *CI* **h.** line *LE* **i.** line *JB*

2. a. How are lines *AE* and *JK* related? Find the slopes of these lines. What do you notice?

b. How are lines *FK* and *HJ* related? Find the slopes of these lines. What do you notice?

3. Find two other pairs of parallel lines. Find the slopes of the lines in each pair. What do you notice? Explain why this is true.

4. **a.** What kind of shape is quadrilateral *HFKJ*? Explain how you know.

   b. Points *J* and *K*, along with two other labeled points, are the vertices of another quadrilateral with the same kind of shape as quadrilateral *HFKJ*. Which points are the other two vertices of this quadrilateral?

5. **a.** If line segments *KL* and *BC* are extended until they meet, what will be true about the angle formed by their intersection?

   b. How are the slopes of lines *KL* and *BC* related?

6. **a.** Find a line perpendicular to line *AJ*. Give the slope of the line you found. How is the slope related to the slope of line *AJ*?

   b. Find a line perpendicular to line *BL*. Give the slope of the line you found. How is the slope related to the slope of line *BL*?

7. **a.** On dot paper, draw a line with slope $\sqrt{8}$. Explain the steps you use to draw your line.

   b. On the same paper, draw a line with a slope that is a rational number slightly less than $\sqrt{8}$, and a line with a slope that is a rational number slightly greater than $\sqrt{8}$. Give the slopes of the lines you draw.

Connections

8. **a.** What is the length of the hypotenuse of this triangle?

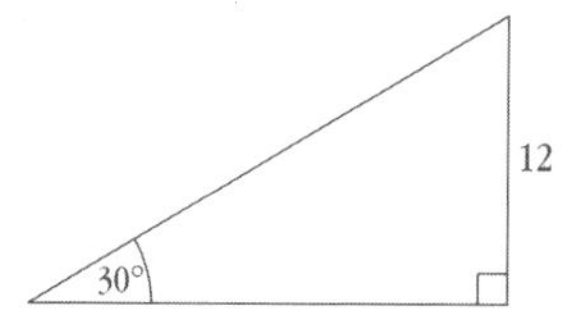

   b. What is the slope of the hypotenuse?

   c. What is the area of the triangle?

4a. *HFKJ* is a parallelogram because both pairs of opposite sides are parallel (the line segments composing pairs of opposite sides have the same slope).

4b. points *L* and *B*

5a. The two lines will form right angles, so the lines are perpendicular.

5b. Line *KL* has slope $^-3$; line *BC* has slope $\frac{1}{3}$. These slopes are negative reciprocals. (Note: Students may use other language to express the relationship between the slopes.)

6a. Line *IB* is perpendicular to line *AJ* and has slope $^-\frac{2}{3}$. This is the negative reciprocal of $\frac{3}{2}$, the slope of line *AJ*.

6b. Line *LC* is perpendicular to line *BL* and has slope $^-1$. This is the negative reciprocal of 1, the slope of line *BL*.

7. See page 72h.

Connections

8a. This is a 30-60-90 triangle, so the hypotenuse is twice the length of the shorter leg, or 24.

8b. Since $24^2 - 12^2 = 432$, the longer leg has length $\sqrt{432}$. The hypotenuse thus has slope $\frac{12}{\sqrt{432}} \approx \frac{12}{20.78} \approx$ 0.58.

8c. The triangle has an area of about $\frac{1}{2} \times 20.78 \times 12 = 124.7$ square units.

9. See page 73.

10. Pairs of opposite sides are parallel (with slopes 1 and $^-1$) and adjacent sides are perpendicular (with slopes that are negative reciprocals), so the figure is a rectangle.

11. Two sides are perpendicular (with slopes $\frac{1}{2}$ and $^-2$, which are negative reciprocals) and thus make a right angle, so the figure is a right triangle.

12. The two sides that look as if they might be perpendicular have slopes $^-\frac{1}{2}$ and $\frac{3}{2}$, which are not negative reciprocals, so the figure is not a right triangle.

13. Pairs of opposite sides are parallel (with slopes $\frac{2}{3}$ and $^-\frac{3}{4}$) and adjacent sides are not perpendicular (with slopes that are not negative reciprocals), so the figure is a nonrectangular parallelogram.

14. Pairs of opposite sides are not parallel (as they have different slopes), so the figure is none of these shapes.

9. a. Draw line segments with as many different slopes as possible by connecting dots on 3-dot-by-3-dot grids. Label each segment with its slope. How many different slopes did you find? How do you know you have found every possible slope?

b. Draw line segments with as many different slopes as possible by connecting dots on 4-dot-by-4-dot grids. Label each segment with its slope. How many different slopes did you find?

c. Draw line segments with as many different slopes as possible by connecting dots on 5-dot-by-5-dot grids. Label each segment with its slope. How many different slopes did you find?

In 10–14, use the slopes of the sides to help you determine whether the figure is a right triangle, a rectangle, a nonrectangular parallelogram, or none of these. Copy the figures onto dot paper if you need to.

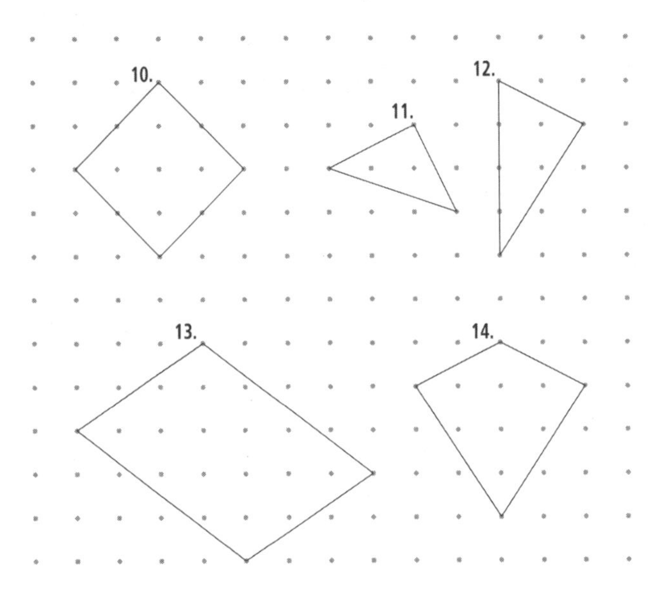

Extensions

15. In the diagram below, line segments connect point P on the circle to the endpoints of diameter AB.

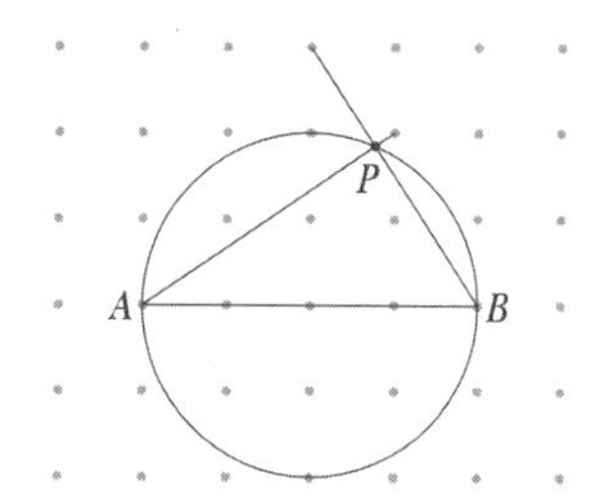

a. Segments PA and PB have each been extended to pass through a second dot. Find the slopes of lines PA and PB. How are the slopes of these lines related? When they intersect, what kind of angle do these lines form?

b. Copy this diagram onto dot paper. Choose another point on the circle, label it Q, and draw segments connecting it to points A and B. Find the slopes of lines QA and QB. How are the slopes of these lines related? What kind of angle do these lines form?

16. a. Find the following ratios of line segment lengths:

$\frac{\text{segment } PP'}{\text{segment } AP'}$ $\frac{\text{segment } QQ'}{\text{segment } AQ'}$ $\frac{\text{segment } RR'}{\text{segment } AR'}$ $\frac{\text{segment } CB}{\text{segment } AB}$

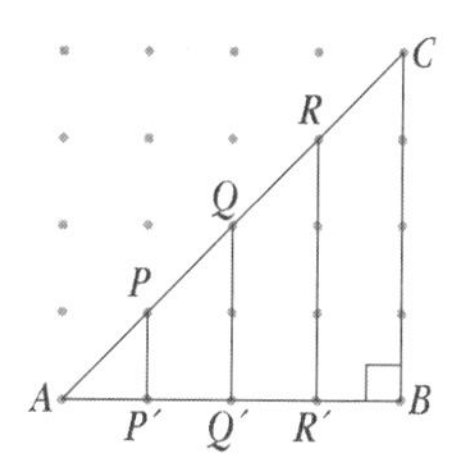

b. What do your answers tell you about the slope of the hypotenuse of triangle ABC?

Extensions

15a. Line PA has slope $\frac{2}{3}$; line PB has slope $-\frac{3}{2}$. These slopes are negative reciprocals, so lines PA and PB are perpendicular and form a right angle.

15b. Any point Q on the circle will result in perpendicular lines AQ and BQ. The slopes of the lines will be negative reciprocals, and the lines will form a right angle. (Note: This helps set the stage for an important theorem in geometry that says any triangle that is inscribed in a semicircle with one side corresponding to the diameter must be a right triangle.)

16a. $\frac{\text{segment } PP'}{\text{segment } AP'} = \frac{1}{1} = 1$,
$\frac{\text{segment } QQ'}{\text{segment } AQ'} = \frac{2}{2} = 1$,
$\frac{\text{segment } RR'}{\text{segment } AR'} = \frac{3}{3} = 1$,
$\frac{\text{segment CB}}{\text{segment } AB} = \frac{4}{4} = 1$

16b. All the ratios show that the slope of the hypotenuse is 1.

Possible Answers

1. We found a line from the origin through a point such that the line had an irrational slope. To do this, we had to find a point that had at least one coordinate that was an irrational number.

2. Two lines are parallel if they have the same slope. And, if two lines have the same slope, they are parallel.

3. Two lines are perpendicular if their slopes are negative reciprocals. And, if the slopes of two lines are negative reciprocals, the lines are perpendicular.

4. To find the distance between two points that do not lie on a vertical or horizontal line, use the line segment between the points as the hypotenuse of a right triangle. Find the squares of the lengths of the legs and add them. The length of the hypotenuse is the square root of this sum. For example, since $3^2 + 1^2 = 10$, the distance between points *A* and *B* below is $\sqrt{10}$.

5. To find the slope of a line through two points, find the ratio of the vertical change to the horizontal change. For example, the slope of the line through points *A* and *B* is $\frac{3}{1}$.

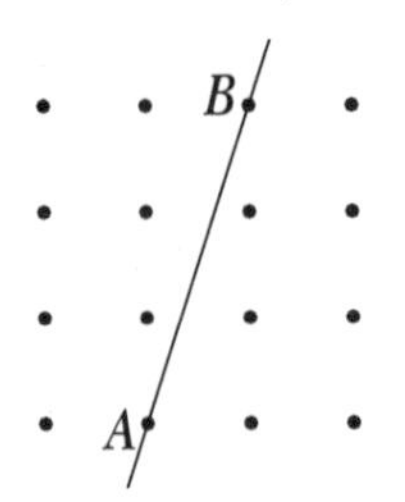

Mathematical Reflections

In this investigation, you explored the slopes of lines, and you drew lines with slopes that are irrational numbers. These questions will help you summarize what you have learned:

1. How did you use slopes to help Oskar escape from the forest?
2. How can you use slopes to determine whether two lines are parallel?
3. How can you use slopes to determine whether two lines are perpendicular?
4. Explain how you would find the distance between two dots on a grid without measuring.
5. Explain how you would find the slope of a line passing through two dots on a grid.

Think about your answers to these questions, discuss your ideas with other students and your teacher, and then write a summary of your findings in your journal.

Tips for the Linguistically Diverse Classroom

Original Rebus The Original Rebus technique is described in detail in *Getting to Know Connected Mathematics.* Students make a copy of the text before it is discussed. During the discussion, they generate their own rebuses for words and phrases they do not understand; the words are made comprehensible through pictures, objects, or demonstrations. Example: Question 1—Key words for which students might make rebuses are *slopes* (/), *Oskar* (stick figure), *forest* (trees).

TEACHING THE INVESTIGATION

6.1 • Revisiting Slopes

In this problem, students are introduced to a computer game in which a character named Oskar moves through a dot grid (a forest of "laser trees"). Oskar must escape from the forest along a straight-line path.

Launch

Talk about the video game Caitlin is playing. Transparency 6.1A shows Oskar's current position in the forest.

> This grid is an immense forest filled with rows of laser trees. Oskar is in the center of the forest. If he bumps into a laser tree, he gets zapped. He has only enough power left to walk in a straight line.
>
> Do you think Caitlin can find a straight-line path to get Oskar out of the forest without his getting zapped by a laser tree?

Transparency 6.1B shows the laser-tree grid labeled with letters and x- and y-axes.

> Can you describe a straight-line path that you think will work?

Encourage students to give pairs of two points, one of which may be the origin, to describe a path. Having students specify the points on a line by giving their coordinates allows you to sketch the familiar triangle associated with slope to review the concept.

(2, 5)
rise
(0, 0) run

$$\text{slope} = \frac{\text{rise}}{\text{run}} = \frac{5}{2}$$

> Giving the slope of a line would be helpful for describing the path Oskar might take. How can you find the slope of a line given the coordinates of two points on the line? *(The slope is the ratio of the change in the vertical direction to the change in the horizontal direction. You can find it by subtracting each pair of coordinates and forming a ratio with the differences.)*

Ask students who think they have paths for Oskar to display them at the overhead.

Here are two paths students might draw:

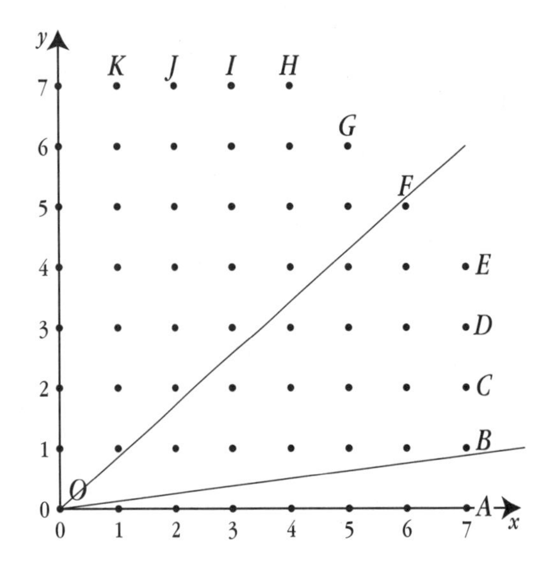

The video game becomes more interesting when students consider that the forest extends beyond the screen: lines that seem to be an escape path for Oskar may be blocked by a laser tree beyond the screen. For example, the lower path shown above just misses the point (7, 1) but would collide with the laser tree at point (8, 1). If the forest ends at the one-thousandth tree along the x-axis, this lower path could be redrawn to exit the forest just below the point (1000, 1).

Have students explore the problem and follow-up in pairs.

Explore

You may need to remind some students of how to read the dot grid as a coordinate grid. Also, make sure that students understand the concept of slope.

Where on this grid is the line with slope $\frac{2}{1}$?

Look at this point. What is the slope of the line from the origin through this point?

Summarize

Have students report the slopes of the lines and explain how they determined them. List the slopes in order from least (line *OA* with slope 0) to greatest (line *OK* with slope 7).

When we look at our list, we see that the slope increases. Does it ever equal 1? *(No, but it comes close; slope 1 would be between $\frac{5}{6}$ and $\frac{6}{5}$.)*

Can you find a line with slope 1? What must be true about the points on this line? *(This can be drawn on the grid. The x- and y-coordinates of the points on this line must be equal.)*

Can you find a line with a slope greater than 7? *(Students would have to extend the grid to find another point in the K column, say L. If L is the point above K, the line OL will slope $\frac{8}{1}$.)*

As the line gets closer and closer to the y-axis, what happens to the slope? *(It increases.)*

As the line approaches a vertical line, the numerator is increasing while the denominator is decreasing, or getting closer to 0. When this happens, the slope increases without bound, or approaches infinity.

The follow-up questions review the relationship between slopes and points on a line. In question 1, students find the coordinates of two more trees on line *OC*.

Is the slope of the line that goes through each of these new points the same as the slope of line *OC*, or has it changed?

Students should recalculate the slope and find that it is the same; this is the essence of a linear relationship.

Will the line that passes through the point $(1, \frac{3}{2})$ hit a tree? *[The slope is $\frac{3}{2}$ over 1, or $\frac{3}{2}$. So, the line will hit the trees at (2, 3), (4, 6), (6, 9), and so on.]*

In question 2, students are asked to find three points on a line with a given slope. In questions 3 and 4, they make the connection between lines that pass through two points and the idea of rational and irrational numbers: the slope of a line is a rational number, or the ratio of two whole numbers, if it passes through two points on the grid.

6.2 • Escaping from the Forest

In this problem, students help Oskar escape from the forest by finding a line that will not hit a tree no matter how far the forest extends. To meet this criteria, students discover, the line must have a slope that is an irrational number.

Launch

Introduce the problem of finding a path out of the forest when the size of the forest is unknown. Then ask:

Suppose Caitlin chooses a straight-line path that corresponds to a line with slope $\frac{278}{997}$. Can she be sure Oskar won't hit a laser tree? *(No, because the forest might extend far enough to include the tree that is to the right 997 units and up 278 units from Oskar.)*

Suppose she chooses a path that corresponds to a line with slope $\frac{1000}{2}$. Can she be sure Oskar won't hit a tree? *(No, because the forest might include the tree that is to the right 2 units and up 1000 units from Oskar.)*

Can you state Caitlin's dilemma using the idea of slope? *(Caitlin needs to find a straight-line path for Oskar that will not hit any trees no matter how far the forest extends. To define such a line, we need to know its slope.)*

Will Oskar's path hit a tree if he walks along a line with slope $\frac{4}{5}$, or $\frac{8}{15}$, or $\frac{43}{900}$? *(All of the lines defined by these slopes will hit a tree if the forest is large enough.)*

Which trees would Oskar hit? *[The trees at point (5, 4), point (15, 8), and point (900, 43), respectively.]*

Students need to understand the idea that no matter what line with a *rational* slope Oskar walks along, he may get zapped by a tree if the forest extends far enough.

Can you describe a line that will not pass through a tree? *(To miss all the trees that are possibly in the forest, the line must pass only through points that do not have whole-number coordinates.)*

So, can Oskar *ever* escape from the forest if it is very, very large? Let's do some more thinking about this problem.

Direct the class's attention to the diagram in Problem 6.2 in the student edition and on Transparency 6.2, which shows the lines with slope $\frac{7}{5}$ and slope $\frac{3}{2}$ (as decimals, 1.4 and 1.5). The lines are quite close together.

Can we specify a line between these two lines that will not hit any of the trees that are shown here?

To find a line between these lines, students will have to find a slope that is between 1.4 and 1.5. If they don't think of it on their own, suggest that they consider the new kinds of numbers that they have met, irrational numbers.

Do you know an irrational number between 1.4 and 1.5?

Once you have elicited the number $\sqrt{2}$, ask how a line with slope $\frac{\sqrt{2}}{1}$ can be drawn.

How can you find a line with slope $\frac{\sqrt{2}}{1}$? Will the line hit a tree? How do you know? Think about these questions as you work on the problem.

Have students work on the problem and follow-up in groups of three or four.

Explore

Students may need help finding a line with slope $\frac{\sqrt{2}}{1}$. They must first locate a point on the vertical axis that corresponds to $\sqrt{2}$. (Their work in Problem 5.1 will help them with this task.) Then, they need to find the point $(1, \sqrt{2})$. A line from the origin through this point will not hit a tree.

As students work, encourage them to find a way to explain *why* a line with an irrational slope will not hit a laser tree no matter how far the forest extends. Diagrams such as the following may help students think about how to solve Caitlin's problem.

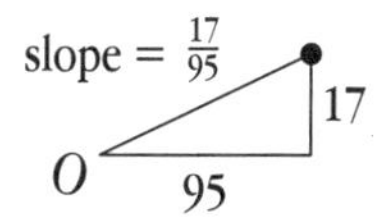

A path with a slope of $\frac{17}{95}$ will hit the tree at (95, 17).

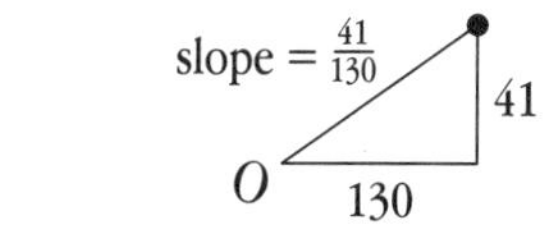

A path with a slope of $\frac{41}{130}$ will hit the tree at (130, 41).

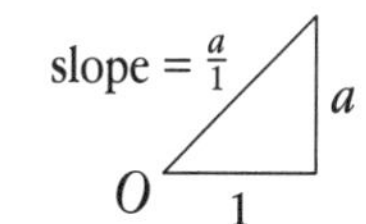

If a is an irrational number, the slope will be irrational. This means the slope is not a ratio of whole numbers, so the line will never hit a tree.

For the Teacher: Finding a Line with Slope $\sqrt{2}$

If Oskar walks along the line with slope $\sqrt{2}$, he can escape from the forest. Here is one way to locate this line.

Draw the line segment from (0, 0) to (1, 1). This line segment has a length of $\sqrt{2}$.

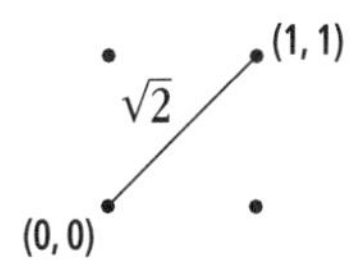

Mark the length of this line segment on a paper strip. Use the paper strip to locate the point (1, $\sqrt{2}$).

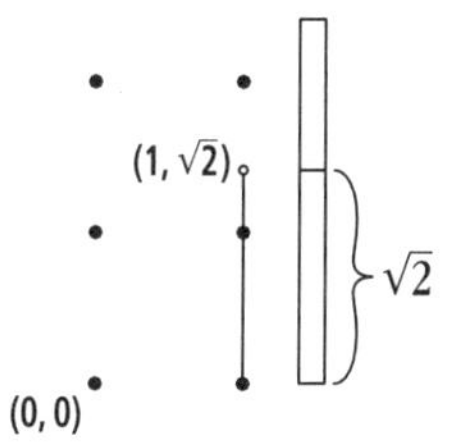

Draw the line through (0, 0) and (1, $\sqrt{2}$). This line has slope $\frac{\sqrt{2}}{1}$ and will never hit a point on the grid with whole-number coordinates.

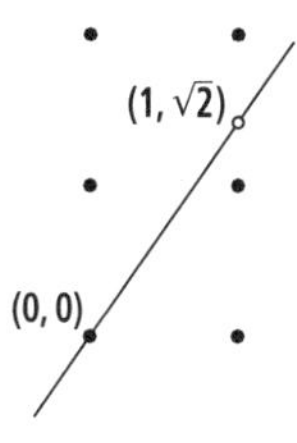

Discuss the follow-up, which gives students a chance to practice finding lines through (0, 0) with other irrational slopes, such as $\sqrt{3}$, $\sqrt{5}$, and $\sqrt{7}$.

Summarize

From their exploration of the problem, students can now indicate a straight-line path for Oskar by describing the slope of the line.

> How did you find a line with slope $\frac{\sqrt{2}}{1}$? *[We found $\sqrt{2}$ on the vertical axis. Then, we found the point (1, 2). The line from the origin through this point has slope $\frac{\sqrt{2}}{1}$.]*

Have someone demonstrate how to draw a line with slope $\sqrt{2}$ on a grid on the board or overhead.

> Will this line ever hit a tree? *(No, because $\sqrt{2}$ is not a fraction or a ratio of whole numbers. If a line hits a tree, its slope must be the ratio of whole numbers.)*

Additional Answers

Answers to Problem 6.2 Follow-Up

1. a. To draw a line with slope $\sqrt{5}$, find a line segment of length $\sqrt{5}$ by recognizing that 5 is the sum of the squares $1^2 + 2^2$. Then, use this length to locate the point $(1, \sqrt{5})$. The line through the origin and the point $(1, \sqrt{5})$ is a line with slope $\sqrt{5}$.

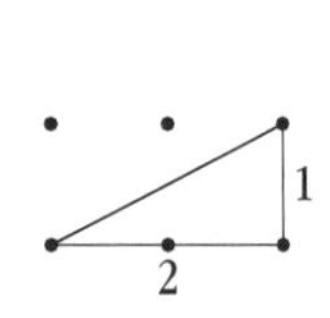

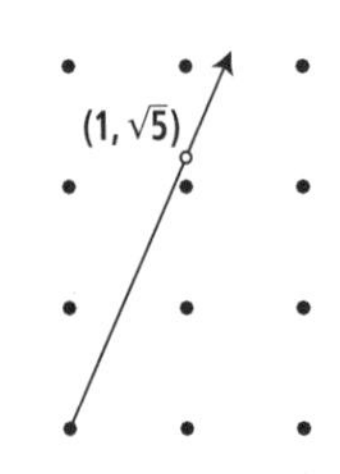

b. Since $\sqrt{5}$ is between 2.2 and 2.3, we can draw lines with slopes 2.2 and 2.3. Since $2.2 = \frac{22}{10}$ and $2.3 = \frac{23}{10}$, we can draw lines from (0, 0) to (10, 22) and from (0, 0) to (10, 23) and get lines on either side of a line through (0, 0) with slope $\sqrt{5}$.

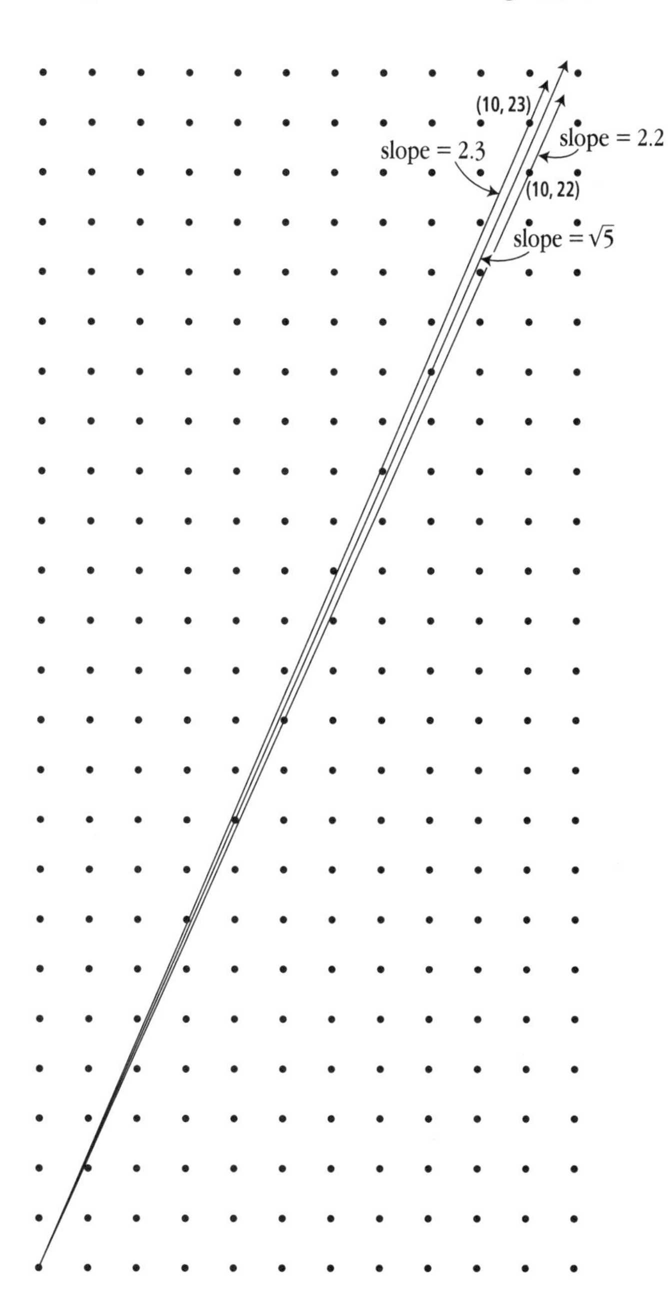

ACE Answers

Applications

7a. Possible answer: I drew lines with slopes $\frac{28}{10}$ and $\frac{29}{10}$. The line with slope $\sqrt{8}$ (or approximately slope $\sqrt{8}$) is located between these lines because $\sqrt{8}$ is between 2.8 and 2.9.

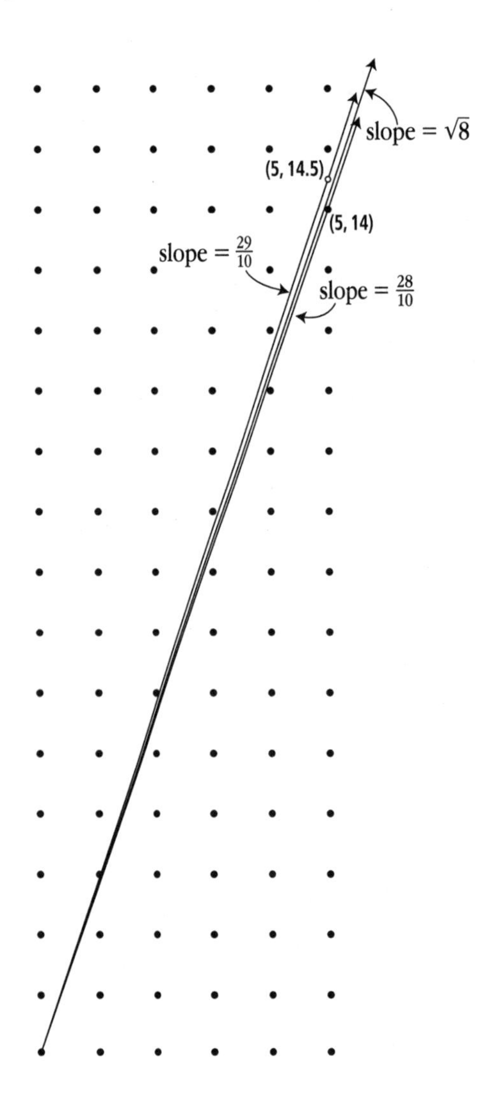

7b. Lines with slopes $\frac{28}{10}$ and $\frac{29}{10}$ are shown in part a.

Connections

9. By starting at the top left corner and drawing segments that go through all the other points, students can find all the negative slopes (as well as slope 0 and the line with no slope). By starting at the bottom left corner, they can find all the positive slopes (being careful not to repeat the horizontal and vertical lines).

9a. Eight slopes are possible.

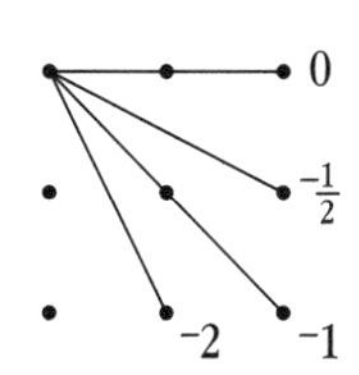

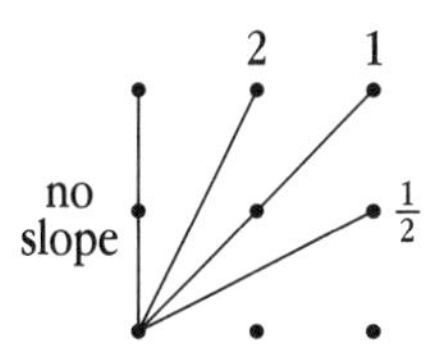

9b. Sixteen slopes are possible.

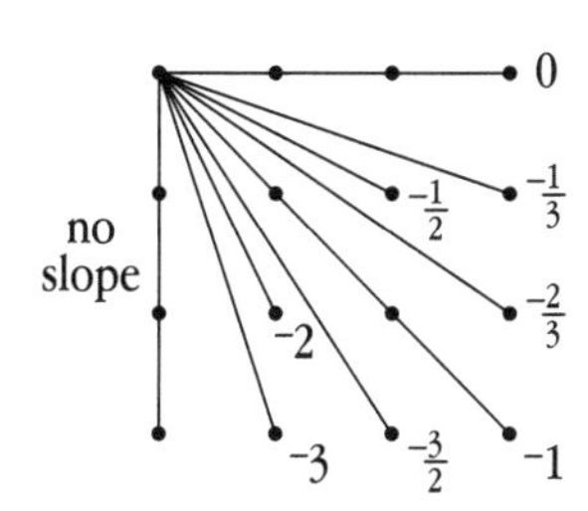

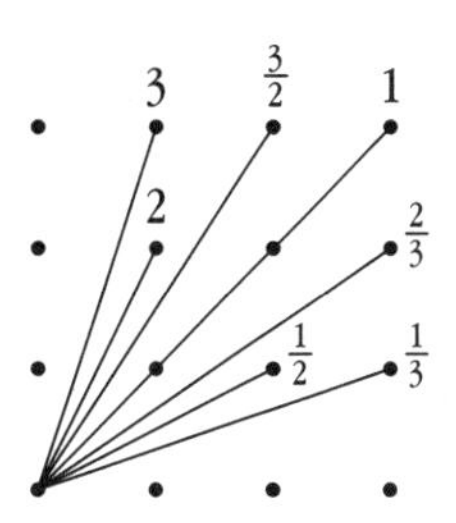

9c. Twenty-four slopes are possible.

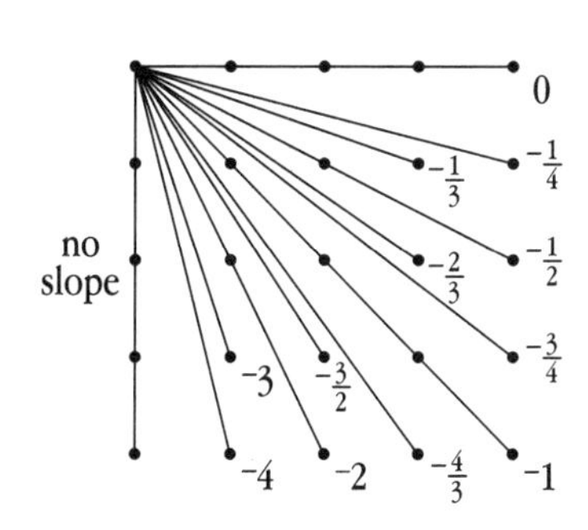

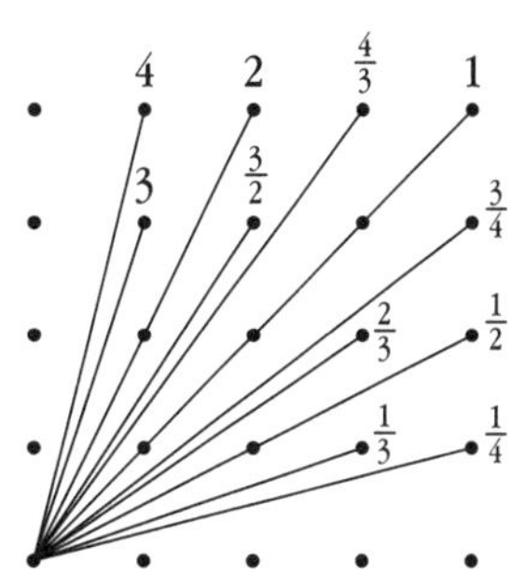

Assessment Resources

Name ______________________ Date ____________

Check-Up

1. *A* and *B* are two vertices of a square.

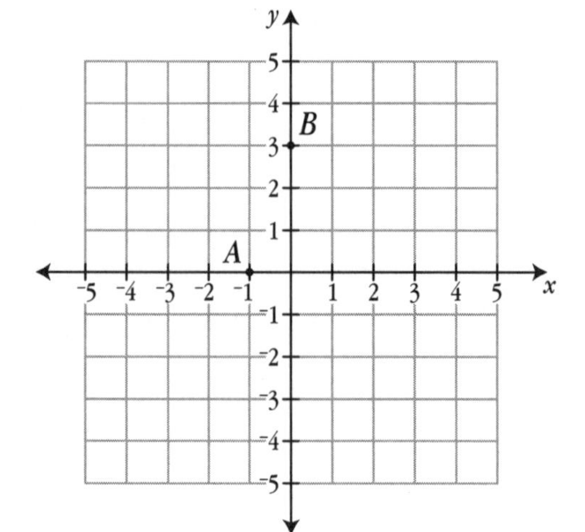

   a. What could the coordinates of the other two vertices be? Add the points to the grid, and label their coordinates.

   b. What is the side length of the square you have identified?

   c. What is the area of the square you have identified?

In 2 and 3, find the area of the polygon. Show all work you do.

2.

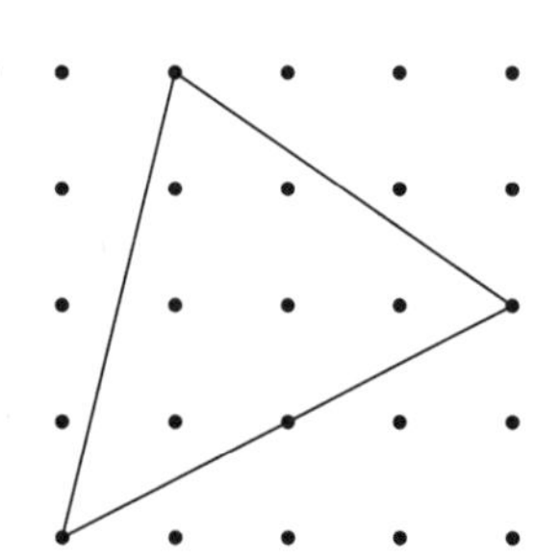

3.

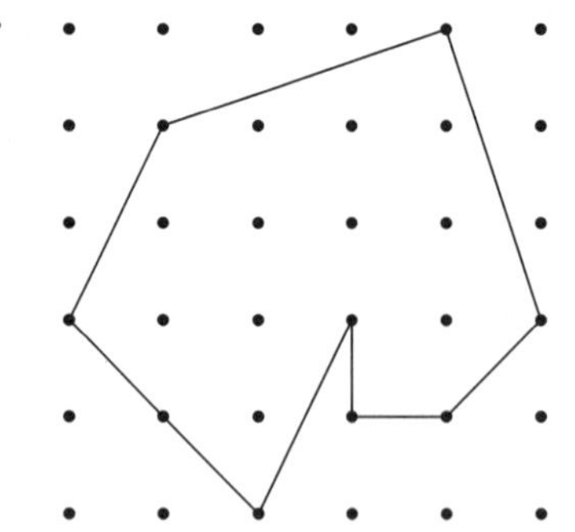

Name ______________________________ Date ______________

Check-Up

4. **a.** On the grid below, identify the point named by each coordinate pair. Connect points *P*, *Q*, and *R* to make a closed figure.

   P (⁻1, ⁻2) *Q* (2, ⁻4) *R* (1, 1)

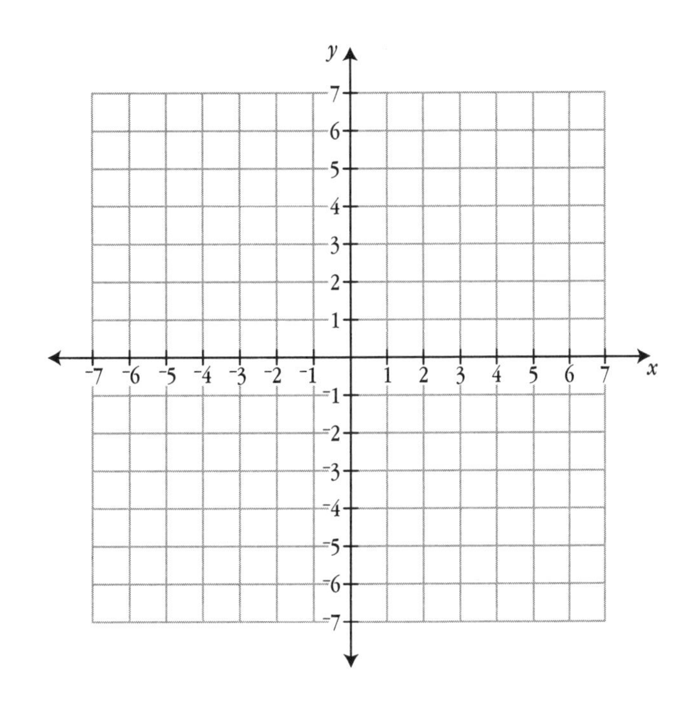

   b. Find the lengths of the sides of figure *PQR* by using areas of squares that match each leg. Show all your work.

   c. What is the area of figure *PQR*?

Names ______________________________ Date ______________

Quiz

1. The three line segments at right are drawn on centimeter dot paper.

   a. Find the length of each segment to the nearest ten-thousandth of a centimeter.

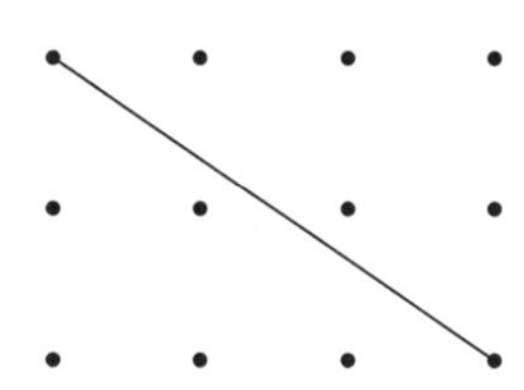

   b. Could these line segments be arranged to form a triangle? If no, explain why not. If yes, answer this question: Could they form a right triangle? Explain why or why not.

2. Use the diagram below to answer parts a–f. Show all work you do to find your solutions.

   a. What is the area of square *ABCD*?

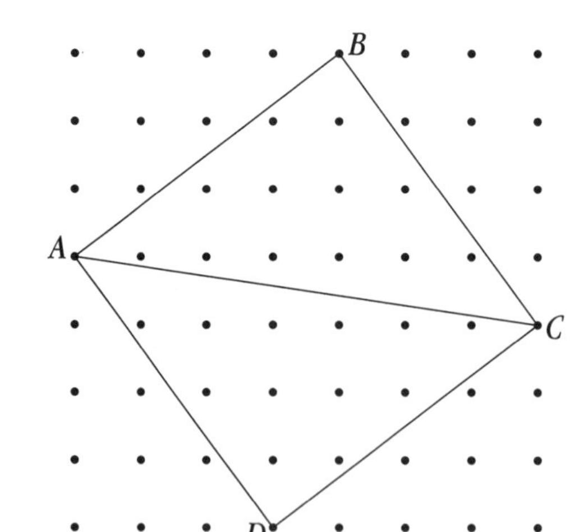

   b. What is the length of line segment *AB*?

   c. What is the distance from *A* to *C*?

   d. What is the area of triangle *ABC*?

Names ______________________ Date ____________

Quiz

e. Explain how the area of square *ABCD* compares to the area of triangle *ABC*.

f. Explain how the perimeter of square *ABCD* compares to the perimeter of triangle *ABC*.

3. The garden gate on the left needs help! The gardener wants to brace the gate by adding a diagonal strip of wood between the two horizontal strips.

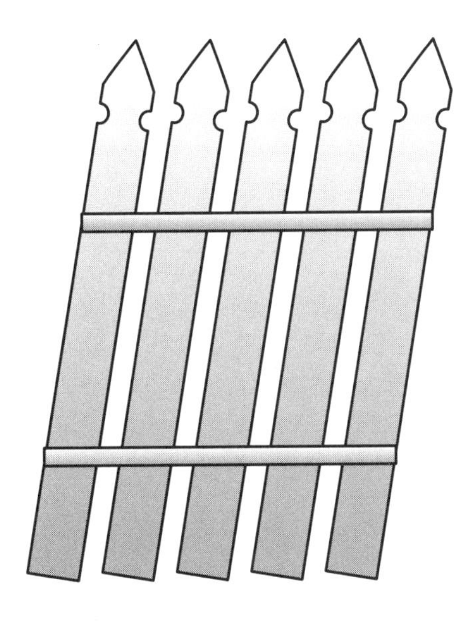

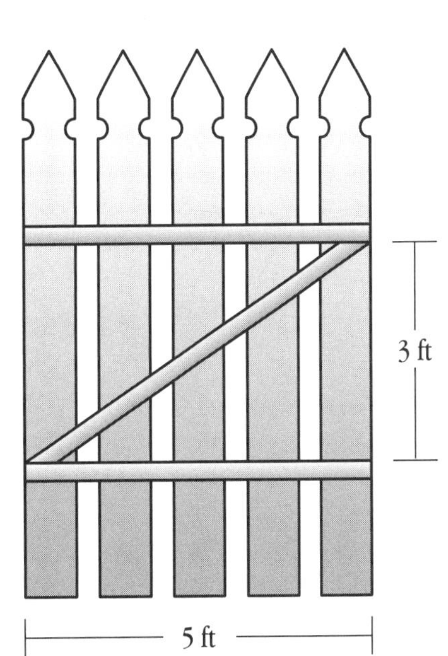

a. How long should the diagonal strip of wood be? (Do not worry about trimming the ends to make a perfect fit.) Show all work you do to find your solution.

b. A standard tape measure is marked in feet and inches. If your answer for part a is written only in feet, rewrite it in feet and inches.

Question Bank

Assign these questions as additional homework, or use them as review, quiz, or test questions.

1. Kathy heard her teacher say that every fraction can be written as a decimal that either repeats or terminates. She tried $\frac{15}{7}$ on her calculator to confirm this, and it gave her an answer of 2.142857143. She was not sure whether this decimal is repeating or terminating, so she worked the problem out by hand as shown. When she reached this point in her division, she said, "*Now* I see that this decimal will repeat." What is Kathy's evidence that this is a repeating decimal?

```
      2.1428571
  7)15.000000000
    14
     10
      7
      30
      28
       20
       14
        60
        56
         40
         35
          50
          49
           10
            7
           30
```

2. **a.** The points *A*, *C*, and *E* are labeled on the grid. Place a fourth vertex, *S*, to form parallelogram *ACES*. Give the coordinates of vertex *S*.

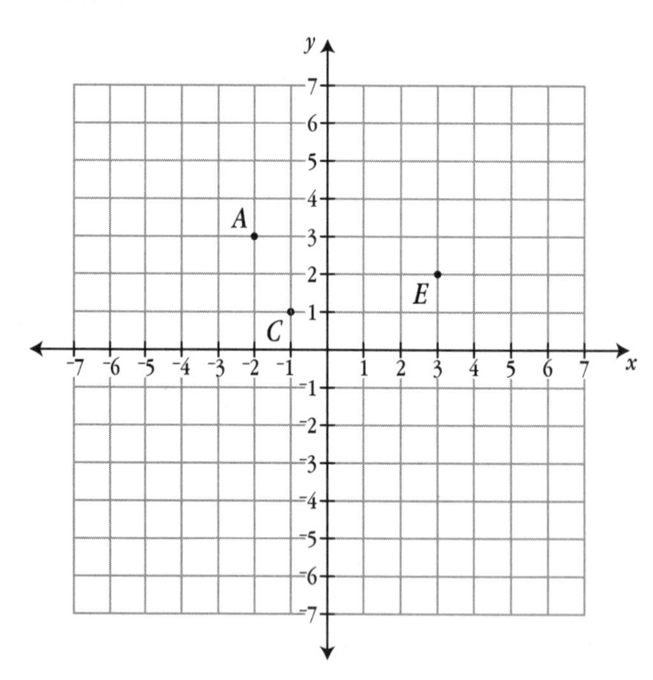

b. Explain how you decided where to put vertex *S*. What facts about the parallelogram helped you to decide on the position?

c. Is there another way to make a parallelogram using points *A*, *C*, and *E* and a fourth vertex? If so, explain where the fourth vertex would be. If not, explain why not.

3. **a.** Find the area of each square below.

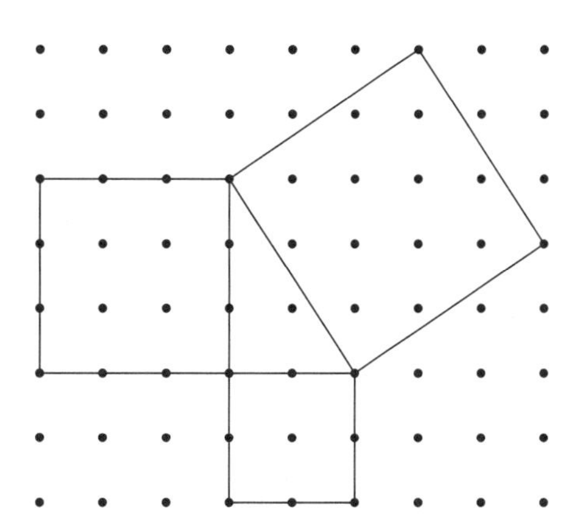

b. Describe any relationship you notice in your answer to part a.

4. Without using a ruler, find the length of segment *AB*.

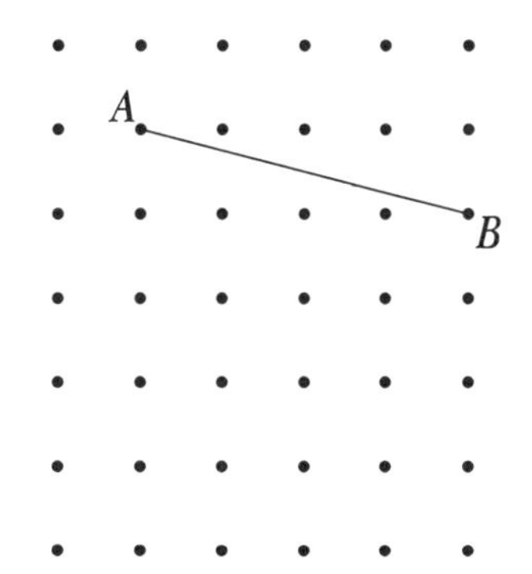

In 5 and 6, use the Pythagorean Theorem to find the length of the line segment. Show all work you do to find your solutions.

5. What is the length of segment *AB*?

6. What is the length of segment *CD*?

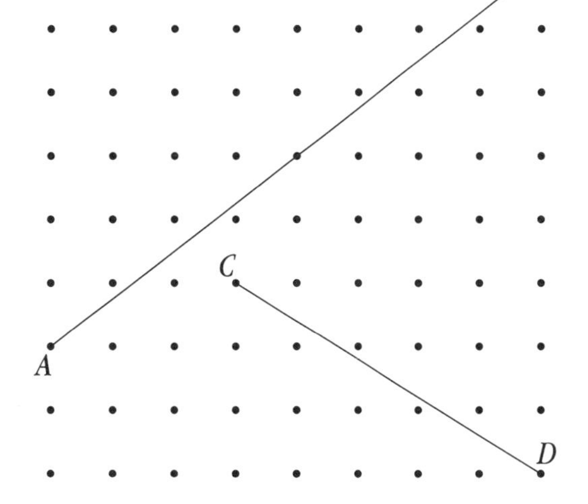

7. Redbeard's treasure is buried at one of the grid points shown below at a distance of $\sqrt{13}$ from the *X*. Where could the treasure be? Explain how you located a possible position.

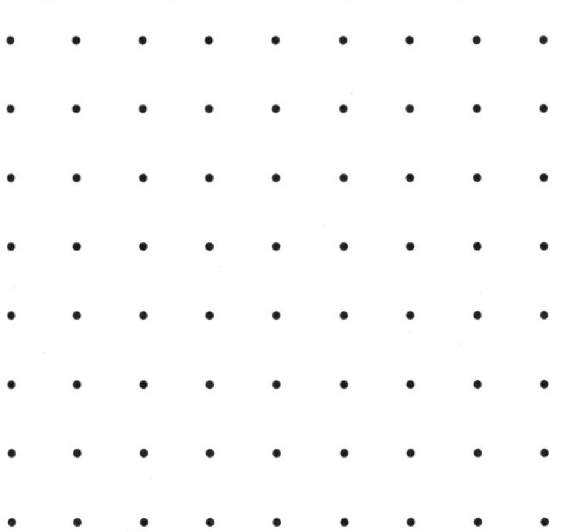

8. Celia's brother says his algebra teacher said that $\sqrt{20}$ is the same as $2 \times \sqrt{5}$, but this does not seem correct to him. Celia made this sketch to explain why this is indeed true. Explain how Celia used her sketch to help her brother understand.

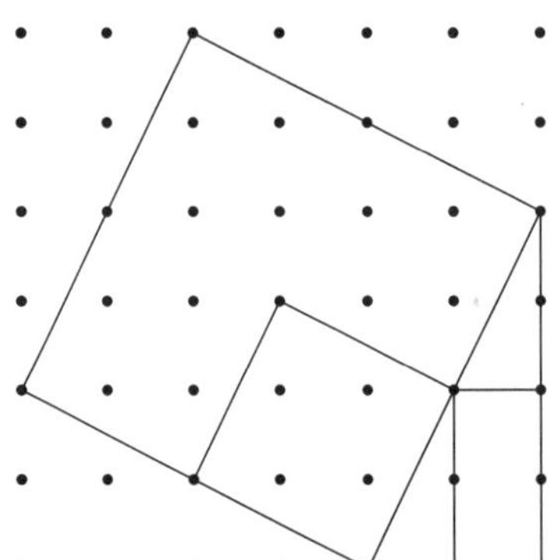

9. Find the slopes of all the line segments on the grid.

slope of line a:_____ slope of line b:_____

slope of line c:_____ slope of line d:_____

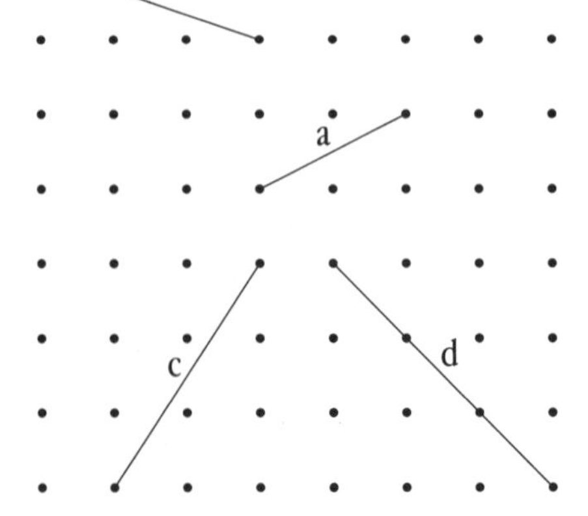

10. **a.** Starting at point A, use a ruler to draw as many line segments as you can, all with different slopes. Each line segment must start at A and end at another grid point. Give the slope of each line segment.

b. How do you know you have drawn all the possible line segments?

11. **a.** Give the coordinates of points A and B.

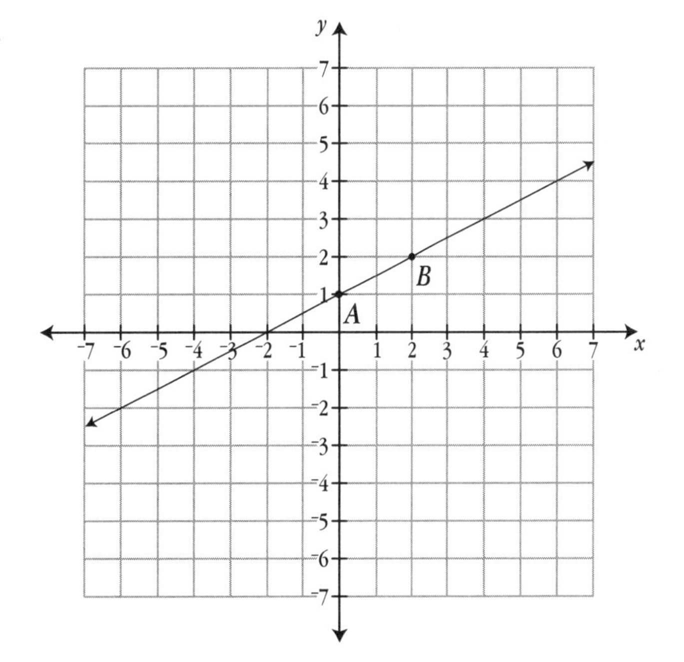

b. What is the slope of line AB?

c. What is an equation of line AB?

12. a. Starting at point A, use a ruler to draw and label line segments on the grid below with slopes 1, $\frac{1}{2}$, $\frac{1}{3}$, $\frac{1}{4}$, and $\frac{1}{5}$. (The first line segment is drawn as an example.)

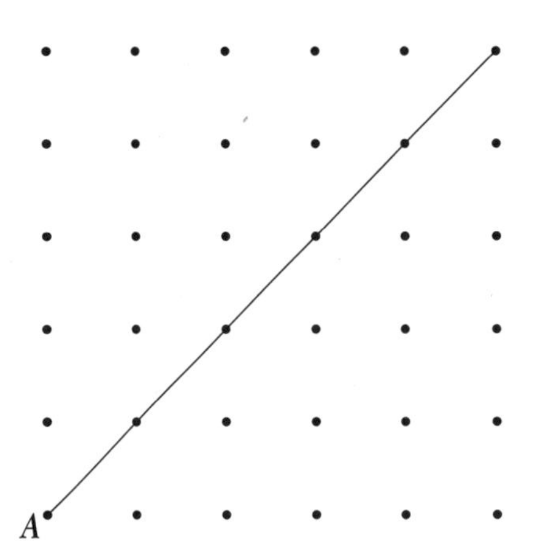

b. Using a different color, draw and label more line segments, all starting at point A, with slopes of $\frac{2}{1}$, $\frac{2}{2}$, $\frac{2}{3}$, $\frac{2}{4}$, and $\frac{2}{5}$.

c. Are any of the line segments you drew in part b the same as the line segments you drew in part a? Explain why this does or doesn't happen.

13. A 14-foot piece of wire is strung between a building and the ground, making a 30-60-90 triangle as shown.

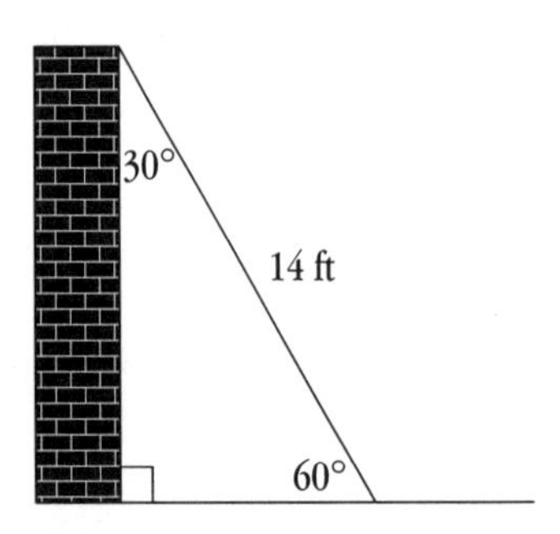

a. How far straight out from the base of the building is the wire attached to the ground?

b. How far up the side of the building is the wire attached?

Name ______________________________ Date ______________

Unit Test

1. Write each fraction as a repeating or terminating decimal. Be sure to indicate whether the decimal repeats or terminates by the way you write it.

   $\frac{5}{2}$ $\qquad$ $\frac{2}{11}$ $\qquad$ $\frac{3}{7}$

2. Draw line segments to connect points *A, C, E,* and *S,* in that order.

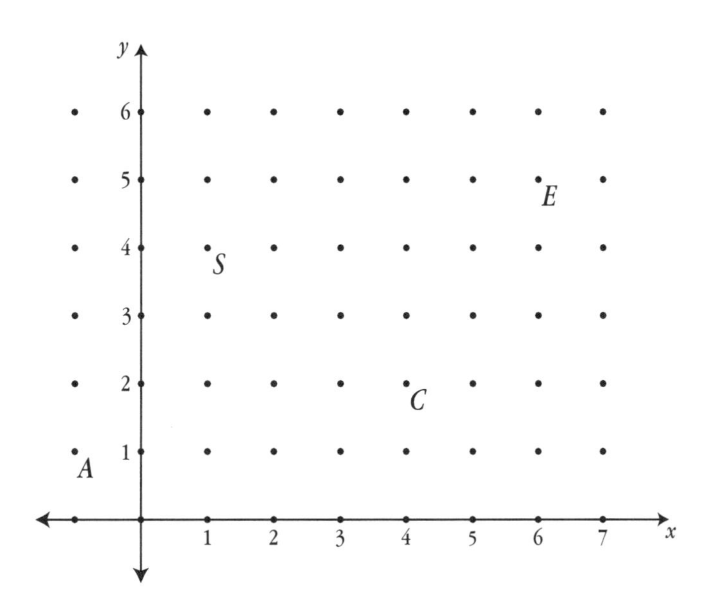

   a. What shape is figure *ACES*?

   b. Without using a ruler, find the lengths of line segments *AS, SE, CE,* and *AC.*

   c. How do the lengths of the sides compare?

   d. Find the slope of each side of the figure: *AS, SE, CE,* and *AC.*

   e. How do the slopes of the sides compare?

   f. What is the area of the figure?

3. Evaluate $\sqrt{50}$. Is this decimal a repeating, terminating, or nonrepeating decimal? Explain how you know.

Name ______________________________ Date ______________

Unit Test

In 4 and 5, find the perimeter and area of the figure.

4.

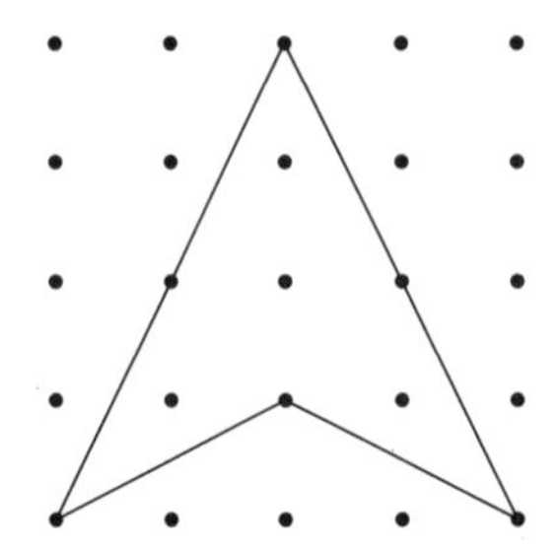

5.

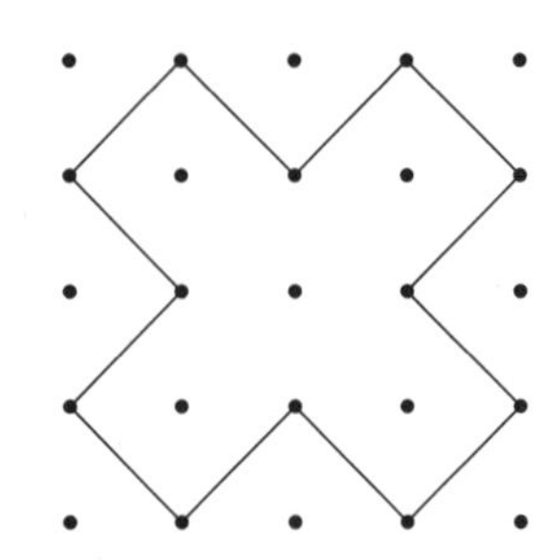

6. **a.** On the dot grid below, draw and label a line segment with length $\sqrt{2}$.

   b. Draw and label a line segment with length $\sqrt{4}$.

   c. Which is greater, $\sqrt{2} + \sqrt{2}$ or $\sqrt{4}$? Explain how you know.

7. Label the grid dot below with the letter T so that the length of ST is $\sqrt{10}$.

Name ______________________________ Date ______________

Notebook Checklist

Journal Organization

_____ Problems and Mathematical Reflections are labeled and dated.

_____ Work is neat and easy to find and follow.

Vocabulary

_____ All words are listed. _____ All words are defined or described.

Check-Up and Quiz

_____ Check-Up

_____ Quiz

Homework Assignments

_____ ______________________________

_____ ______________________________

_____ ______________________________

_____ ______________________________

_____ ______________________________

_____ ______________________________

_____ ______________________________

_____ ______________________________

_____ ______________________________

_____ ______________________________

_____ ______________________________

_____ ______________________________

_____ ______________________________

_____ ______________________________

_____ ______________________________

Name ______________________________ Date ______________

Self-Assessment

Vocabulary

Of the vocabulary words I defined or described in my journal, the word ______________ best demonstrates my ability to give a clear definition or description.

Of the vocabulary words I defined or described in my journal, the word ______________ best demonstrates my ability to use an example to help explain or describe an idea.

Mathematical Ideas

We often need to know distances or lengths. Sometimes we can determine distance by measuring or using a scale on a map. In *Looking for Pythagoras,* I learned to find the distance between two points a new way.

1. **a.** I have learned the following about finding locations, slopes, and areas on a grid; using the Pythagorean Theorem to find areas and lengths; and repeating and nonrepeating decimals:

b. Here are page numbers of journal entries that give evidence of what I have learned, along with descriptions of what each entry shows:

2. **a.** These are the mathematical ideas I am still struggling with:

b. This is why I think these ideas are difficult for me:

c. Here are page numbers of journal entries that give evidence of what I am struggling with, along with descriptions of what each entry shows:

Class Participation

I contributed to the class discussion and understanding of *Looking for Pythagoras* when I . . .
(Give examples.)

Answers to the Check-Up

1. a. If AB is a side of the square, the other points could be (3, 2) and (2, $^-1$) or ($^-4$, 1) and ($^-3$, 4). If AB is a diagonal of the square, the other points are ($^-2$, 2) and (1, 1).

   b. If AB is a side, $\sqrt{10}$. If AB is a diagonal, $\sqrt{5}$.

   c. If AB is a side, 10 square units. If AB is a diagonal, 5 square units.

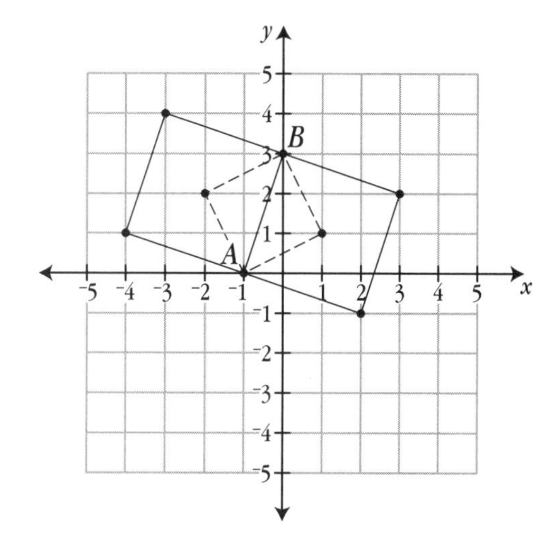

2. 7 square units; Possible strategy: Enclose the triangle in a square, and subtract the area of the three right triangles from the area of the square: 16 – (2 + 3 + 4) = 7 square units.

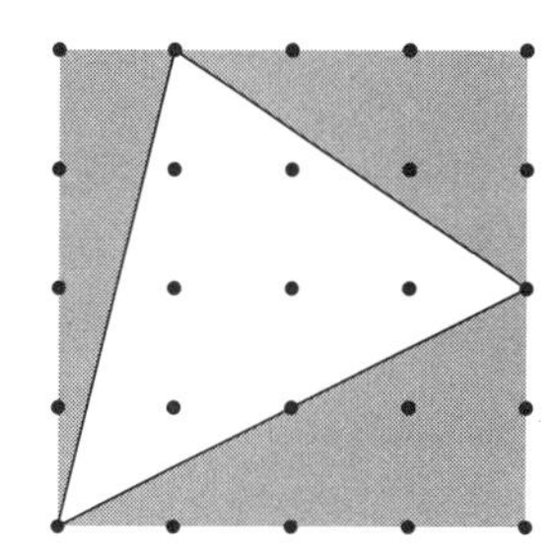

3. 14.5 square units; Possible strategy: Subdivide the figure and add the smaller areas: 1 + 6 + 1.5 + 2 + 1 + 3 = 14.5 square units.

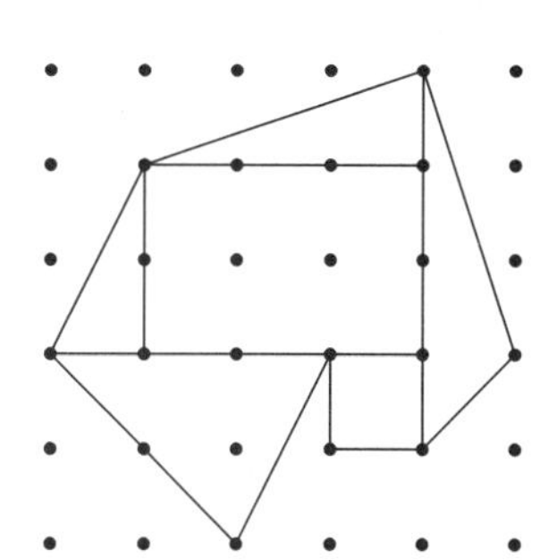

4. **a, b.** The areas of the squares on sides *PR* and *PQ* are both 13 square units, so *PR* and *PQ* both have length $\sqrt{13}$. The area of the third square is 26 square units, so *RQ* has length $\sqrt{26}$.

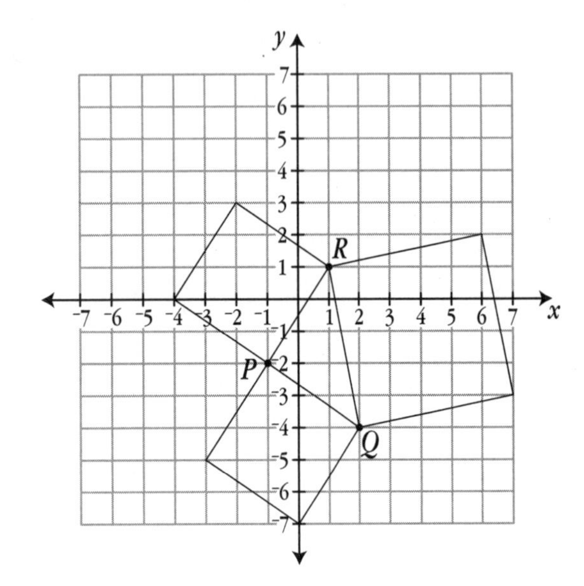

c. $\frac{1}{2}(\sqrt{13})(\sqrt{13}) = 6.5$ square units

Answers to the Quiz

1. **a.**

2.0000

$\sqrt{5} \approx 2.2361$

$\sqrt{13} \approx 3.6056$

b. The three segments will form a triangle because the sum of the two shorter lengths is greater than the longest length: $2 + 2.2361 > 3.6056$. However, they do not form a right triangle because they do not fit the Pythagorean Theorem—that is, $2^2 + (\sqrt{5})^2 \neq (\sqrt{13})^2$.

2. **a.** 25 square units (Students' methods will vary.)

b. The length of segment *AB* is $\sqrt{25}$, or 5.

c. Since $5^2 + 5^2 = 50$, the distance between *A* and *C* is $\sqrt{50}$.

d. The area of triangle *ABC* is half the area of square *ABCD*, or 12.5 square units.

e. The area of the square is twice the area of the triangle.

f. The perimeter of the square is $4 \times 5 = 20$. The perimeter of the triangle is $5 + 5 + \sqrt{50} \approx 17$. The perimeter of the square is slightly greater than the perimeter of the triangle.

3. a. Since $3^2 + 5^2 = 34$, the length of the diagonal strip should be $\sqrt{34} \approx 5.8$ ft.

   b. To convert 0.8 ft to inches, multiply by 12. The result is 9.6, so the strip should be about 5 ft 10 in long.

Answers to the Question Bank

1. Kathy can claim that since the process will continue to "bring down a 0" each time, the string of remainders will cycle through a series of numbers. In the division that is shown, the remainders (with 0s attached) are 10, 30, 20, 60, 40, and 50. They will continue to display this pattern, causing the digits in the decimal to repeat the pattern 142857. (Kathy's calculator included a 3 after the second 4 due to rounding.)

2. a. Vertex *S* is at point (2, 4).

   b. Answers will vary. Students may talk about drawing parallel lines (from the definition of a parallelogram), one through point *A* parallel to segment *CE* and one through point *E* parallel to segment *AC*. These lines meet at the fourth vertex of the parallelogram. Students may also reason using what they know about slopes of parallel lines. For example, the line through points *C* and *E* has slope $\frac{1}{4}$, so the line through points *A* and *S* must also have slope $\frac{1}{4}$. Moving from point *A* up 1 and to the right 4, point *S* is located at (2, 4).

   c. Placing a vertex at point (4, 0) will form a second parallelogram.

3. a. 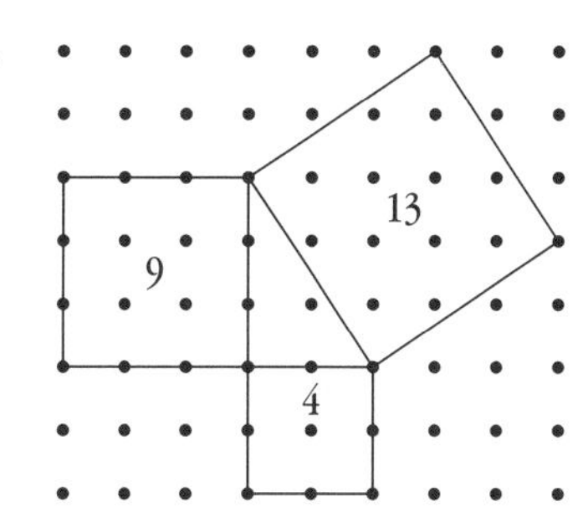

   b. The sum of the areas of the two smaller squares is equal to the area of the largest square.

4. The length of segment *AB* is $\sqrt{17}$. Students might draw a square with segment *AB* as a side, or they might draw a right triangle with segment *AB* as the hypotenuse.

5. Since $6^2 + 8^2 = 100$, the length of segment *AB* is $\sqrt{100} = 10$.

6. Since $5^2 + 3^2 = 34$, the length of segment $CD = \sqrt{34} \approx 5.83$.

7. One way that students may think about this problem is by finding two perfect squares that have a sum of 13 (only 4 and 9 work) and using their square roots (2 and 3) as the vertical and horizontal distances from point *X*. There are three possible locations of the treasure as marked on this map.

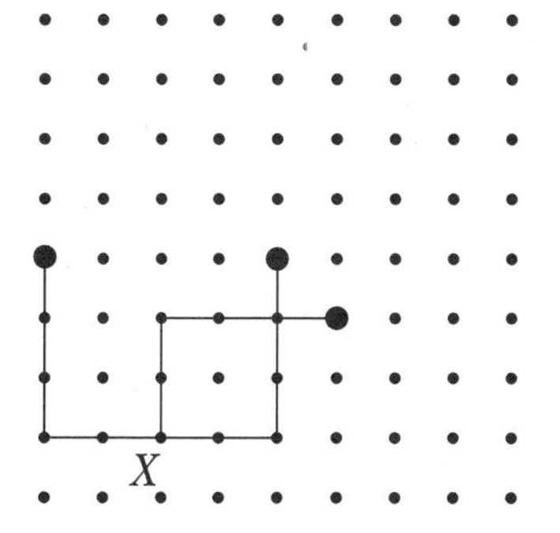

8. Celia's sketch shows that one of the smaller triangles has a hypotenuse of length $\sqrt{5}$ (because the smaller square has an area of 5). The larger triangle has a hypotenuse that is twice as long as that of the smaller triangle (or 2 times $\sqrt{5}$) and it has length $\sqrt{20}$ (because the larger square has an area of 20). So, $\sqrt{20} = 2 \times \sqrt{5}$.

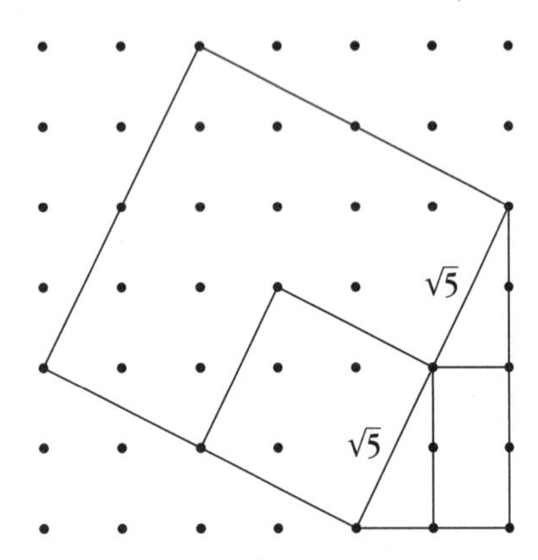

9. slope of line a = $\frac{1}{2}$, slope of line b = $-\frac{1}{3}$, slope of line c = $\frac{3}{2}$, slope of line d = $-\frac{3}{3} = -1$

10. a.

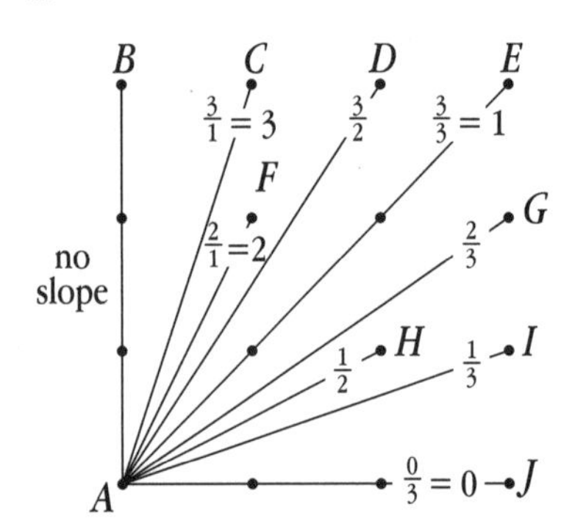

b. Students may say that all of the points are included in one of the line segments, so all the possible slopes have been accounted for. They may also point out that all the combinations of using a 0, 1, 2, or 3 in both the numerator and the denominator of the fractions for the slopes are included (when looking at equivalent fractions such as $\frac{3}{3} = \frac{2}{2} = \frac{1}{1} = 1$).

11. a. Point A is at (0, 1); point B is at (2, 2).

b. The slope of line AB is $\frac{1}{2}$.

c. $y = \frac{1}{2}x + 1$

12. a, b.

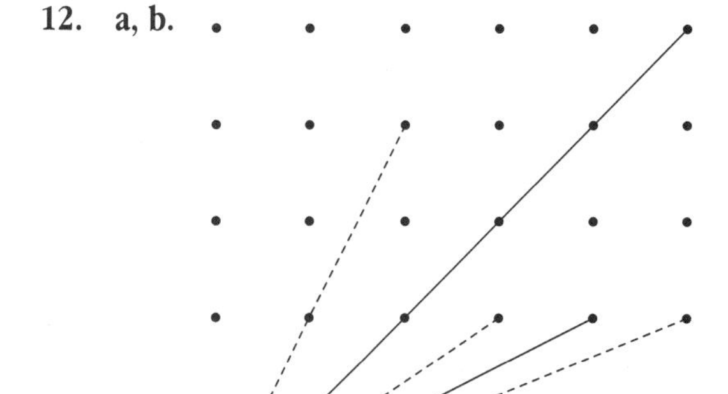

c. Two pairs of line segments are overlying—the segments with slope 1 or $\frac{2}{2}$, and the segments with slope $\frac{1}{2}$ or $\frac{2}{4}$. The slopes are equal, so the line segments lie on the same path.

13. a. Since this is a 30-60-90 triangle, the length of the shorter leg is half the length of the hypotenuse. The wire will be attached to the ground 7 ft from the building.

b. Using the Pythagorean Theorem, since $14^2 - 7^2 = 147$, the wire will be attached $\sqrt{147} \approx 12.12$ ft up the side of the building.

Answers to the Unit Test

1. $\frac{5}{2} = 2.5$, $\frac{2}{11} = 0.1818\ldots$, $\frac{3}{7} = 0.428571428571\ldots$

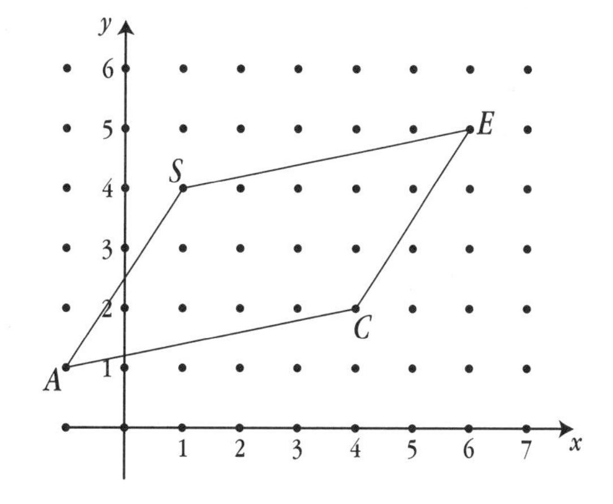

2. a. Figure *ACES* is a parallelogram.

b. The lengths are as follows: *AS*, $\sqrt{13}$; *SE*, $\sqrt{26}$; *CE*, $\sqrt{13}$; *AC*, $\sqrt{26}$.

c. Opposite sides are the same length.

d. The slopes are as follows: *AS*, $\frac{3}{2}$; *SE*, $\frac{1}{5}$; *CE*, $\frac{3}{2}$; *AC*, $\frac{1}{5}$.

e. Slopes of opposite sides are equal.

f. 13 square units

3. The decimal 7.0710678 . . . is nonrepeating. Explanations may include justifications such as taking the calculator output and squaring it and finding it is not exactly 50, so the output is not a terminating decimal. Students may say that they do not see a pattern repeating. They may assert that because 50 is between 49 and 64, which have exact square roots of 7 and 8, respectively, there is no other integer that can be the square root of 50. Any reasonable explanation that reflects that students are using ideas discussed in class is acceptable.

4. area = 6 square units, perimeter = $2\sqrt{20} + 2\sqrt{5} \approx 13.4$

5. area = 10 square units, perimeter = $12\sqrt{2} \approx 17.0$

6. a, b.

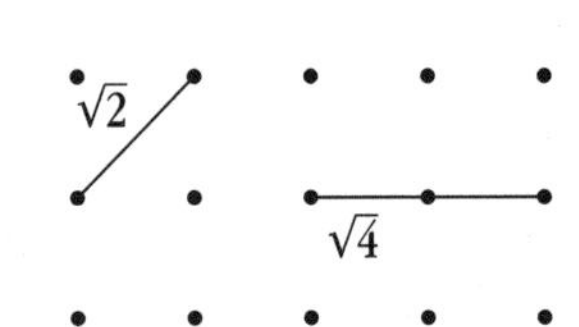

c. $\sqrt{2}$ must be greater than 1 but less than 2, as $1^2 = 1$ and $2^2 = 4$. So, $\sqrt{2} + \sqrt{2}$ must be greater than 2 and therefore greater than $\sqrt{4}$. Also, two line segments of length $\sqrt{2}$ will be longer than a line segment of length $\sqrt{4}$.

7. There are two possible positions:

Guide to the Self-Assessment

Using the self-assessment in each unit, students can reflect on the mathematics in the unit and write about what they have made sense of, what they are still struggling with, and how they contributed to the class's understanding of the mathematics in the unit. The sample below shows how one student approached the self-assessment in this unit. Other students' work using this same assessment tool can be found in the grade 7 units *Stretching and Shrinking* and *What Do You Expect?* It may be helpful to you to read the student papers in those units.

Brad

1a. I learned that you could use the Pythagoras Theorem to find the area of a shape on grid paper and to find lengths of lines. It can help you find the length of sides of a right triangle because if you add the 2 areas together and find the square root of that number, that's the other side of the triangle. I also learned that irrational numbers, like $\sqrt{2}$, cannot be written as a repeating or terminating decimal. Irrational numbers will help Oskar out of the woods because an irrational number can't be written as a fraction. This means that there is no rise over run, or trees to get in the way.

b. Look at journal page 28 and 29. These show that I know how to use the Pythagorus Theorem to solve problems.

2a. I'm struggling with how slopes can be irrational numbers.

b. I don't get it because irrational numbers are lengths of sides of some squares that are drawn with slanted lines on grid paper and so how can they be slopes of lines that don't touch points because these touch points.

c. Page 30 and 31 show that I don't get slopes of irrational numbers.

I contributed when I helped the class see that sevenths were repeating decimals when they thought they weren't.

A Teacher's Comments

Brad's self-assessment is somewhat vague and incomplete. This is typical of the self-assessments I received for this unit from many students in my class. He discusses using the Pythagorean Theorem to find the hypotenuse of a right triangle but fails to address how the theorem can be used to find the length of *any* side of a right triangle if the lengths of the other two sides are known. He also does not elaborate on how right triangles can be formed to determine the length of an oblique, or tilted, line segment on grid or dot paper. We spent a lot of time working on these two ideas in this unit and they should be reflected in this assessment.

Brad writes about repeating and terminating decimals as if he is writing to someone who knows all about these ideas and therefore feels it is not necessary to explain them. The connection he makes between Oskar escaping from the woods and irrational numbers shows that he has some understanding of the relationship between these ideas, but he is somewhat confused about irrational numbers and what it means for a slope to be irrational.

Because Brad's work is typical of the class's work, I will take a class period and revisit this assessment piece by having students give examples of responses to each part of the self-assessment. I will put each question on the overhead and ask the class to tell all that they wrote. We will record these ideas and discuss what additional ideas might be included to give a more detailed and informative response, in effect modeling my expectations for the self-assessment. I will try to address common misconceptions that students showed in their papers. At the end of this process, I will hand back the self-assessments and have each student revise his or her work using ideas from the class discussion. I hope that by using our class-generated ideas about responses as a model, students will better understand what it means to give a complete and thoughtful answer.

I am struck by the lack of detail and completeness in my students' work. I remind myself that this is only the second unit of grade 8 and the start of a new year. I know I must help students understand my expectations and how to use these assessment tools. But the students in my classes were in the Connected Mathematics program in grades 6 and 7. The mathematics teachers in my building are all using the Connected Mathematics self-assessments, and yet at the beginning of every year we must readdress the issue of what it means to give a complete answer.

Because Brad's paper is representative of those of many of my students, I feel I need to do something more. I have asked all the mathematics teachers in our building to attend a meeting on this issue, bringing four examples of student work using the self-assessment component from the most recently completed Connected Mathematics unit. At the meeting, we will all read the examples and discuss strategies for improving the work students do on self-assessments and how we can send consistent messages to them about our expectations for this kind of work every year. I will ask the language arts teachers to attend and to give us suggestions for addressing this problem and others we have with students' written explanations.

Blackline Masters

Labsheets

Transparencies

Additional Resources

Maps of Euclid

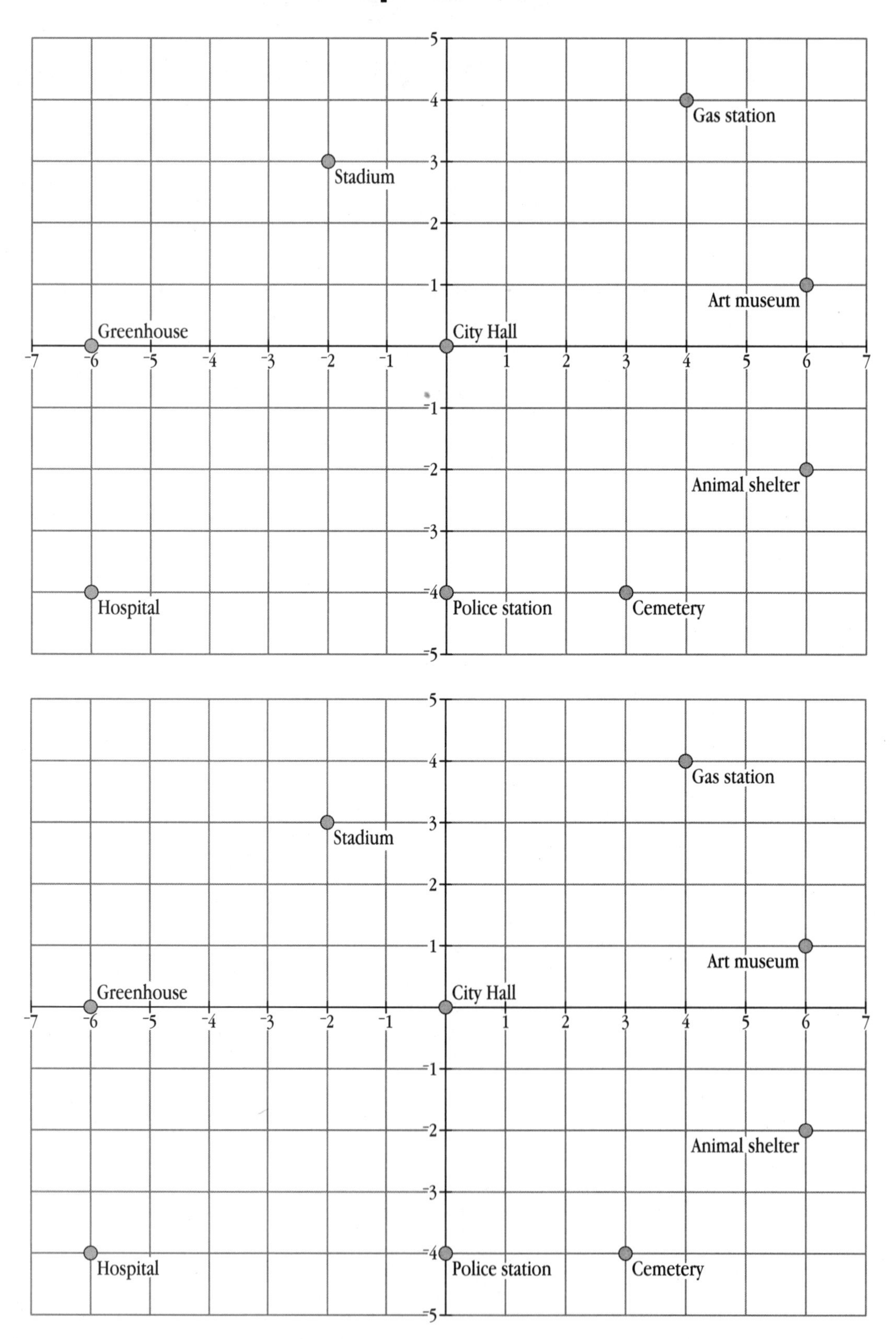

Figures for Problem 2.1 and Follow-Up

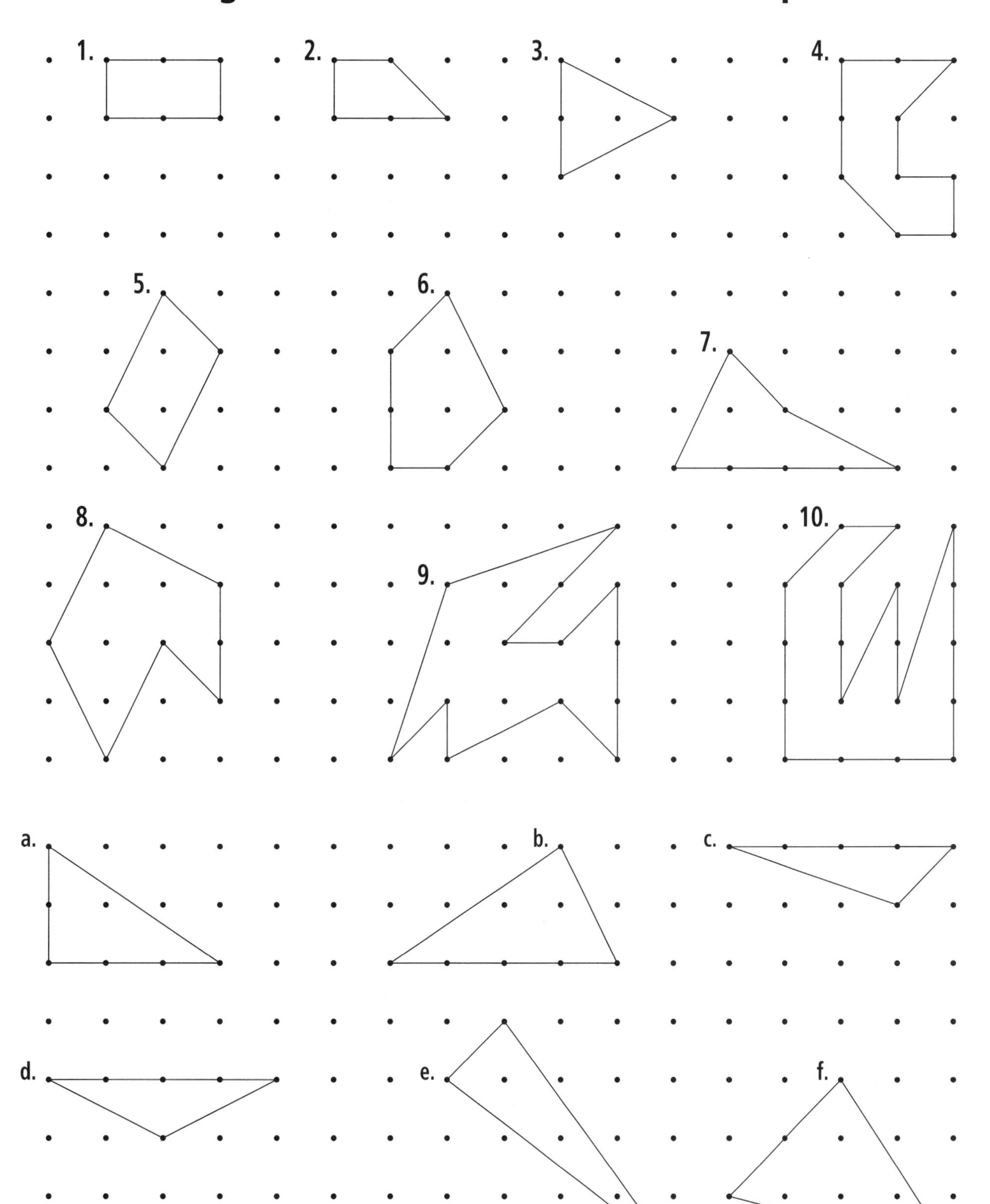

5-Dot-by-5-Dot Grids

Enclosed 5-Dot-by-5-Dot Grids

ACE Questions 1, 2, and 11

Questions 1 and 2

Question 11

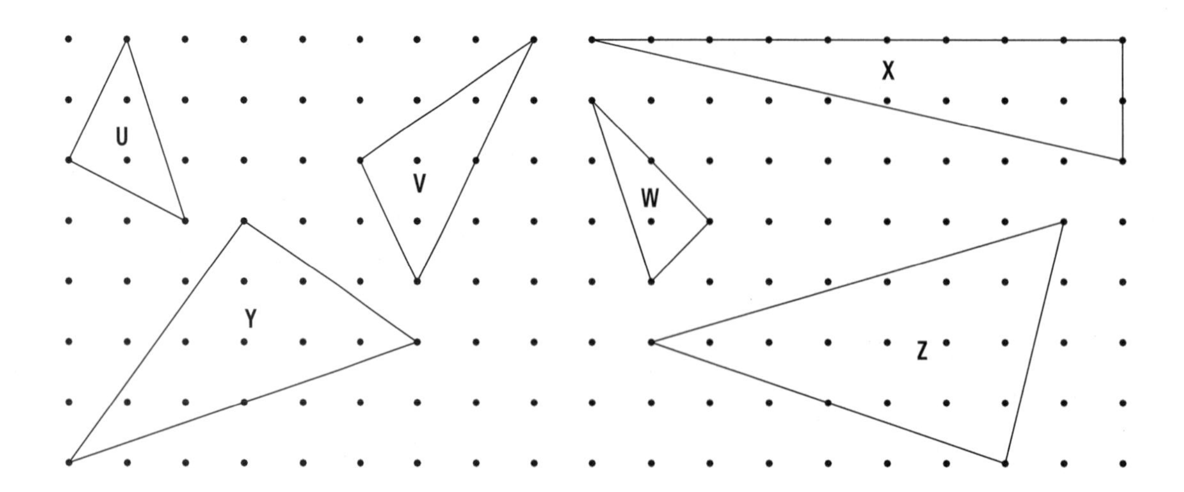

Puzzle Frames and Puzzle Pieces, Set A

Cut out the puzzle pieces. Arrange them to fit in the puzzle frames.

Puzzle frames

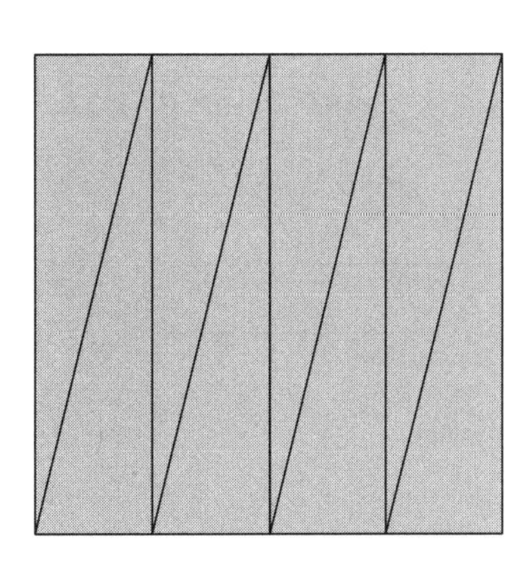

Puzzle pieces

Puzzle Frames and Puzzle Pieces, Set B

Cut out the puzzle pieces. Arrange them to fit in the puzzle frames.

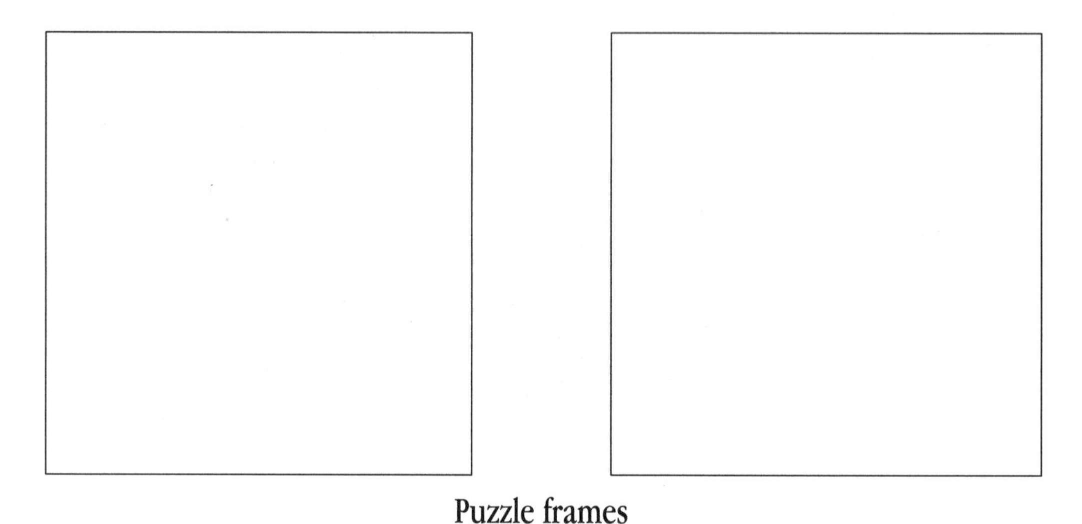

Puzzle frames

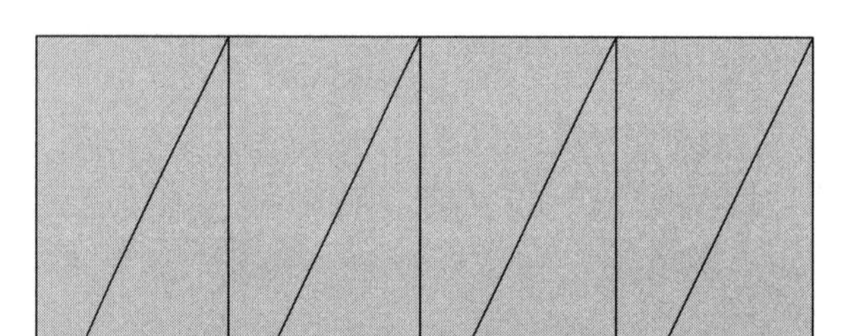

Puzzle pieces

Puzzle Frames and Puzzle Pieces, Set C

Cut out the puzzle pieces. Arrange them to fit in the puzzle frames.

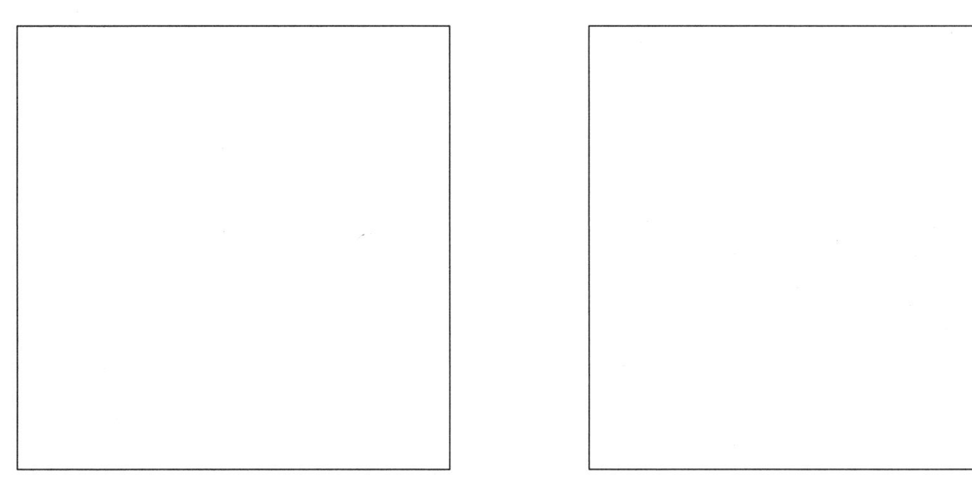

Puzzle frames

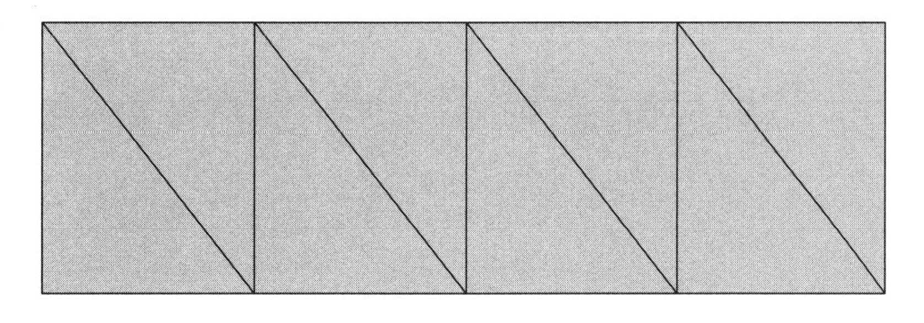

Puzzle pieces

Points on a Grid

B, *D*, *A*, *C*, *F*, *E*

Triangle and Square

Each side of equilateral triangle *ABC* has a length of 2.

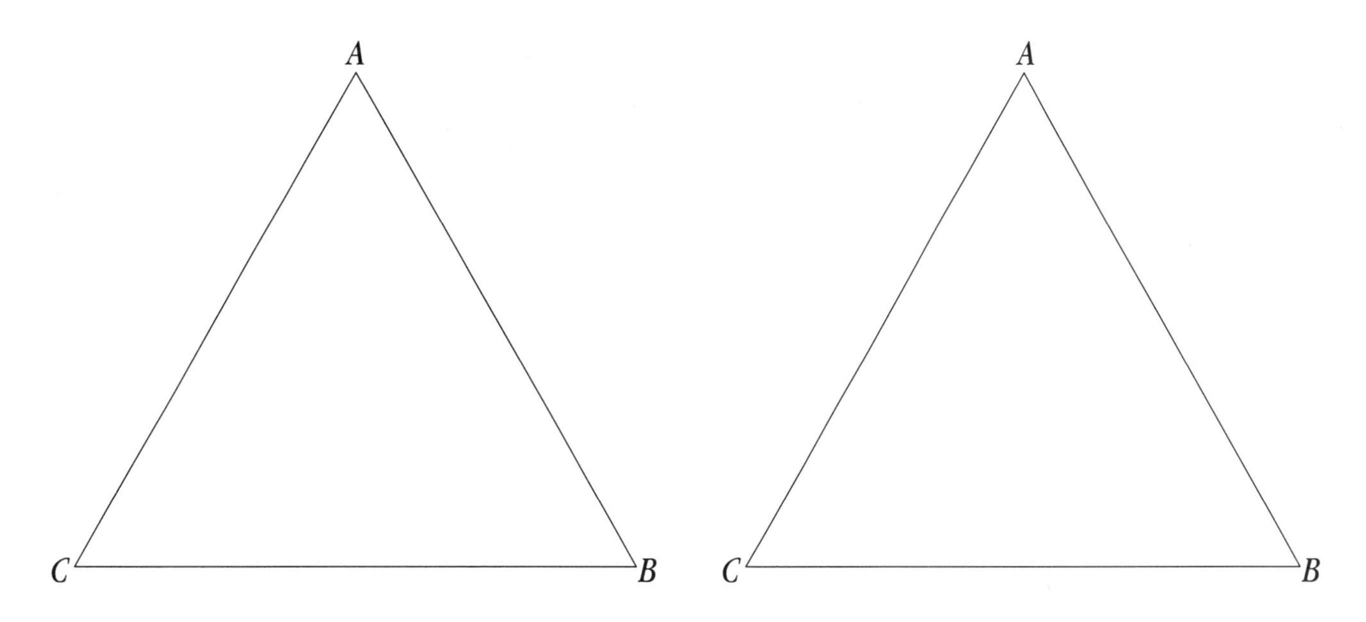

Square *ABCD* has side lengths of 1.

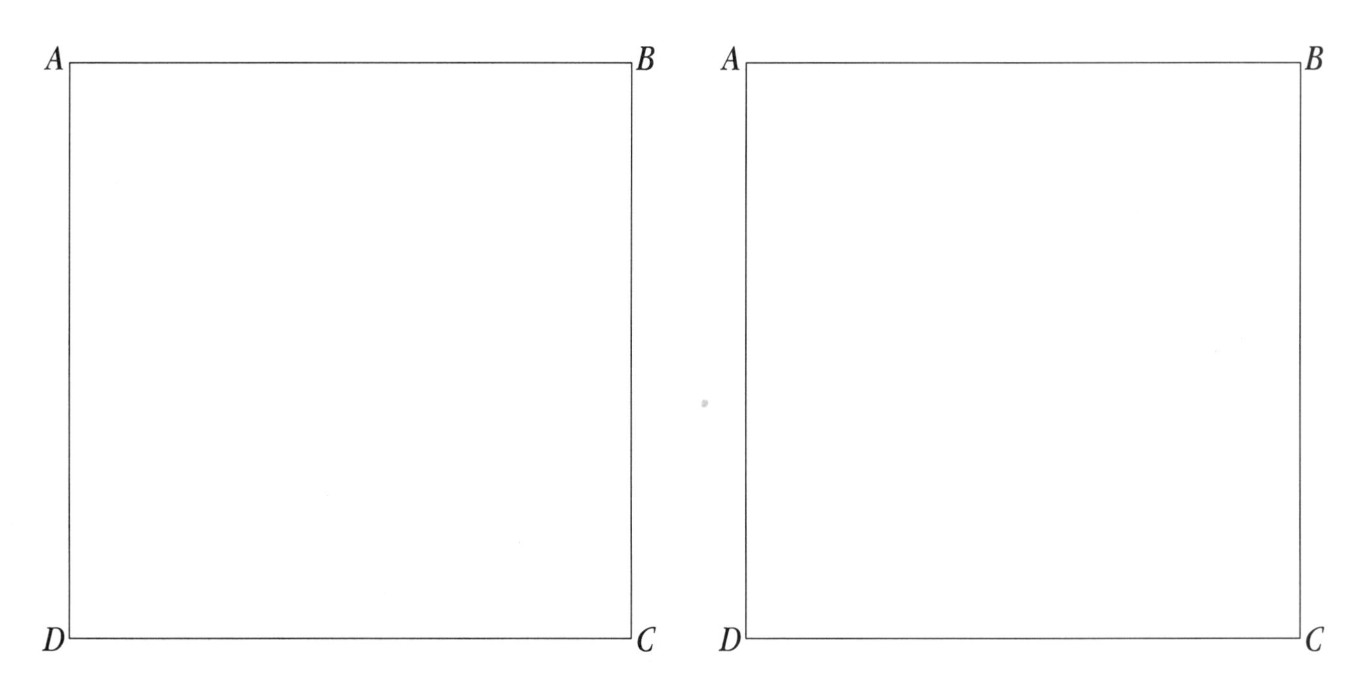

Labsheet 5.1

The Wheel of Theodorus

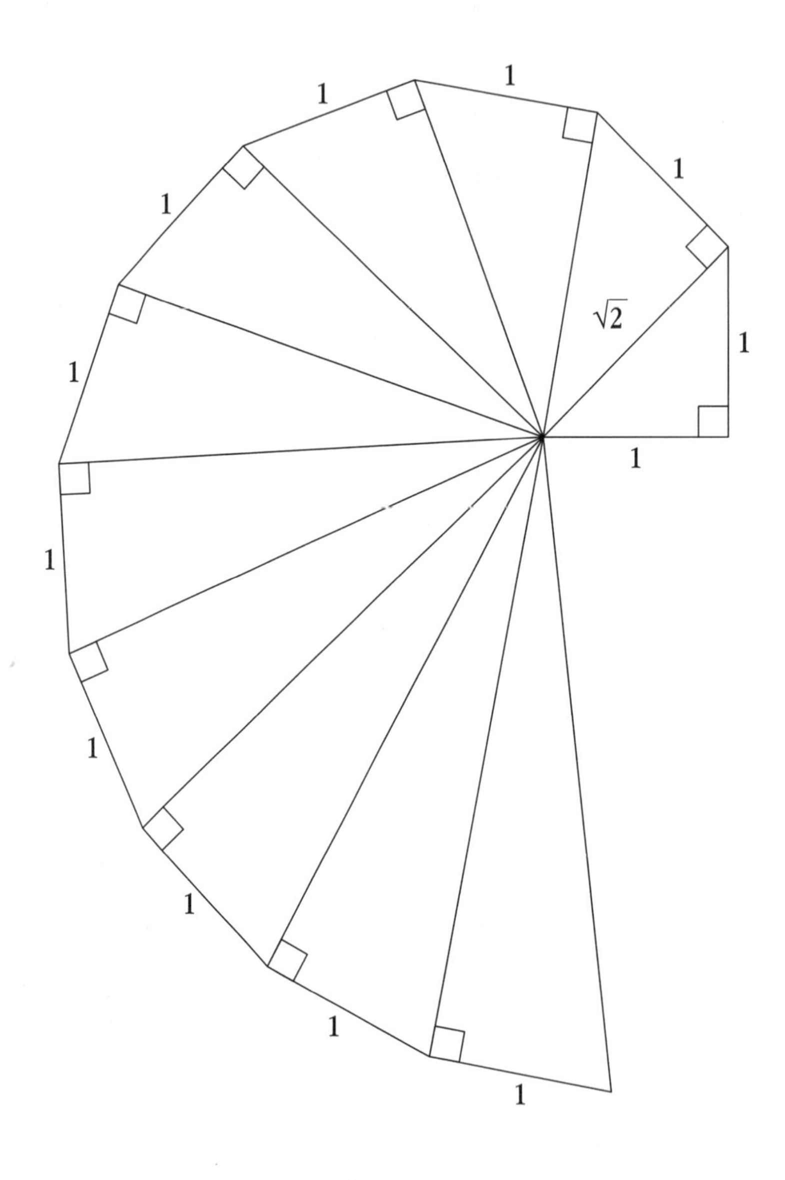

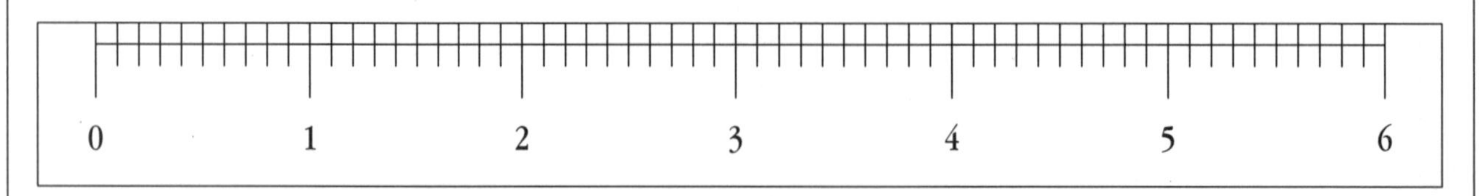

Escaping from the Forest

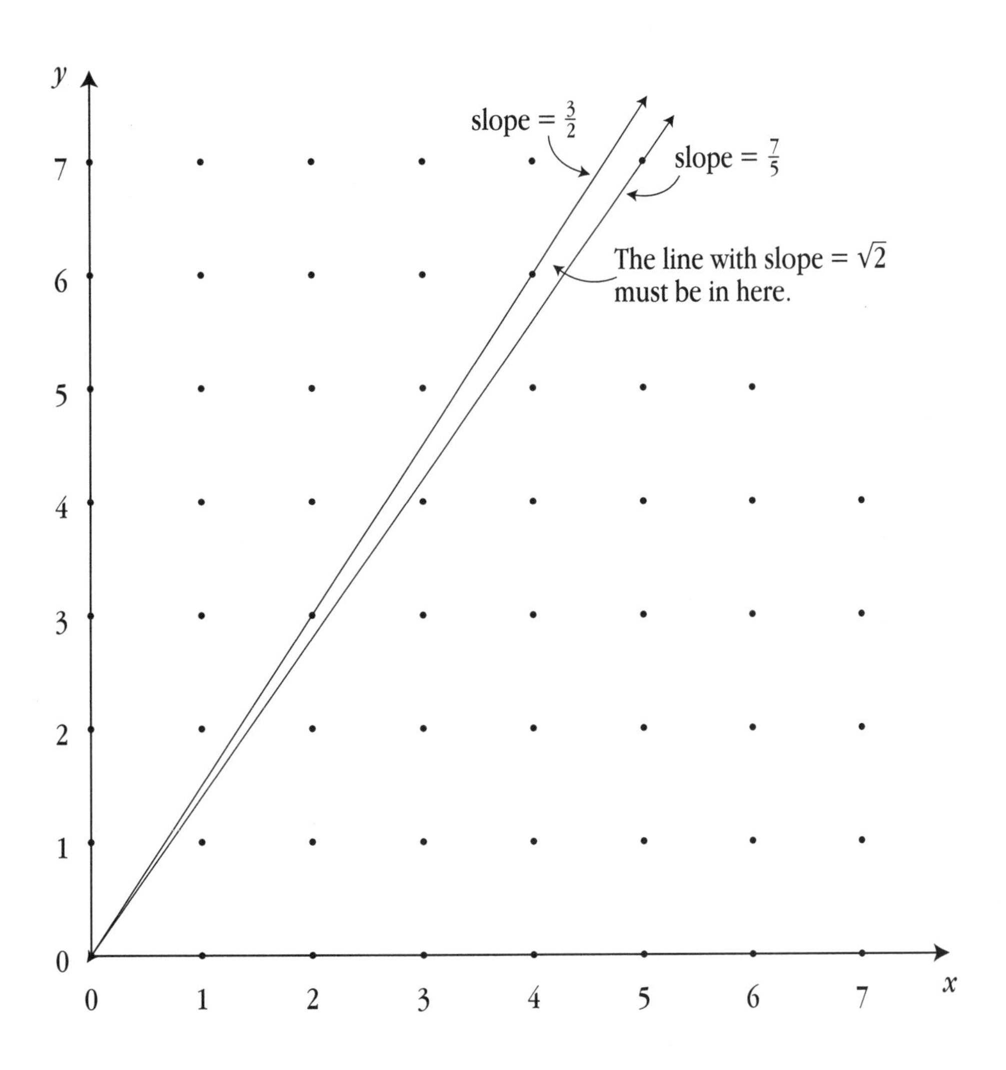

ACE Questions 1–6 and 15

Questions 1–6

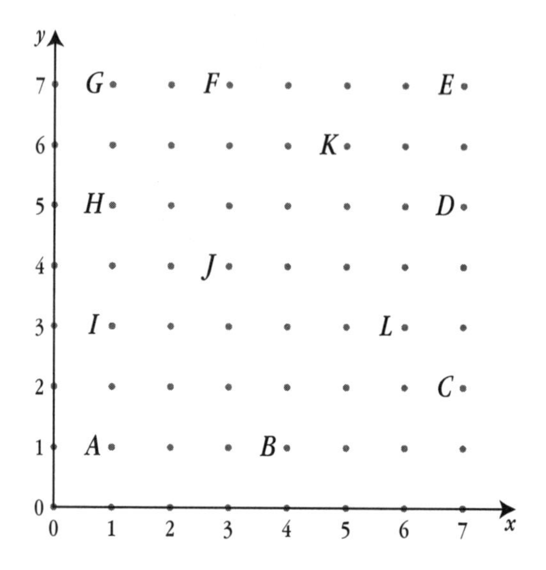

Question 15

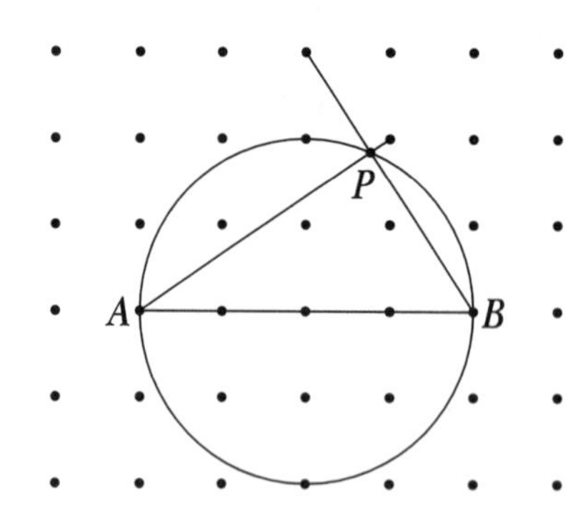

ACE Question 10

Transparency 1.1A

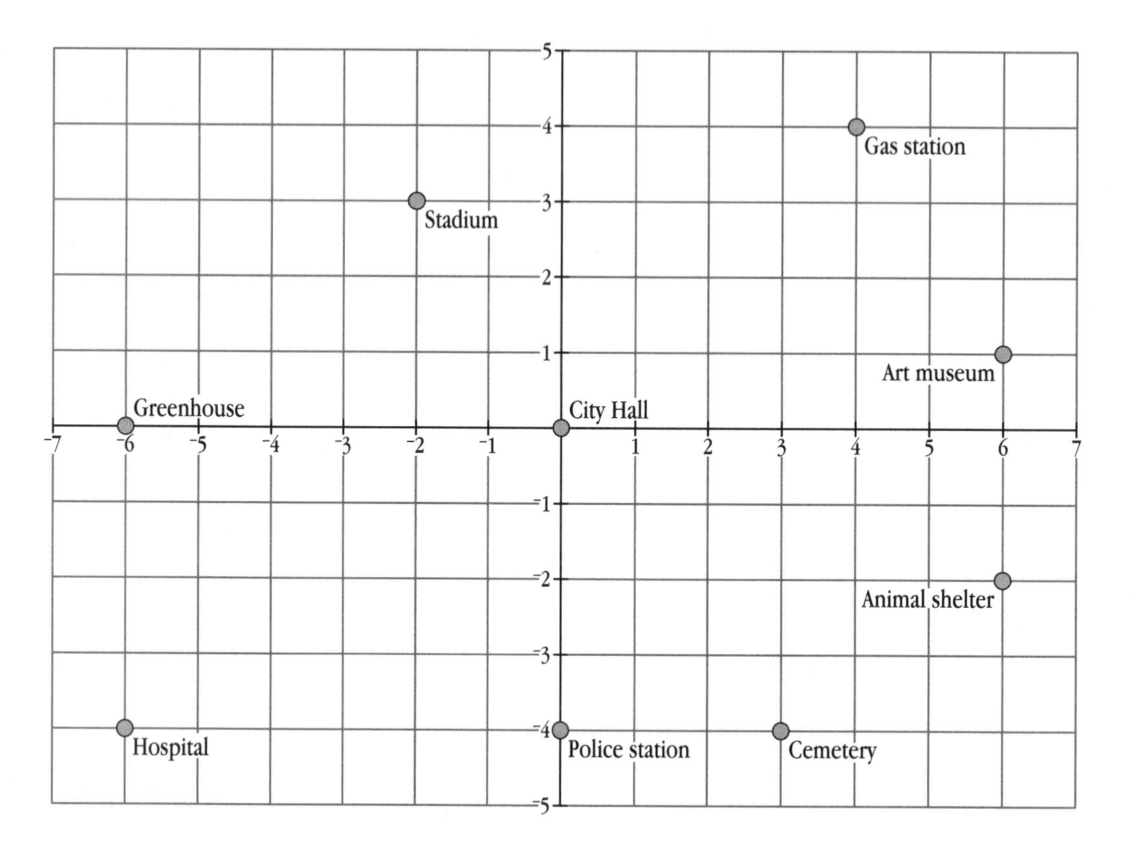

A. Give the coordinates of each labeled landmark on the map.

B. **1.** How many blocks would a car have to travel to get from the hospital to the cemetery?

2. How many blocks would a car have to travel to get from City Hall to the police station?

3. How many blocks would a car have to travel to get from the art museum to the gas station?

C. How can you tell the distance in blocks between two points if you know the coordinates of the points?

Your plan must include the shortest routes between the following pairs of locations. Answer parts A and B for each pair.

Pair 1: the police station to City Hall

Pair 2: the hospital to City Hall

Pair 3: the hospital to the art museum

Pair 4: the police station to the stadium

A. 1. Give the coordinates of each location, and give precise directions for an emergency car route from the starting location to the ending location.

2. Find the total distance, in blocks, a police car would have to travel to get from the starting location to the ending location along your route.

B. A helicopter can travel directly from one point to another. Find the total distance, in blocks, a helicopter would have to travel to get from the starting location to the ending location. You may find it helpful to use a centimeter ruler (1 centimeter = 1 block).

A. If the park with the given vertices is to be a square, what could the coordinates of the other two vertices be? Give two answers.

B. If this park is to be a nonsquare rectangle, what could the coordinates of the other two vertices be? Give two answers.

C. If this park is to be a right triangle, what could the coordinates of the other vertex be? Give two answers.

D. If this park is to be a parallelogram that is *not* a rectangle, what could the coordinates of the other two vertices be? Give two answers.

A. Find the area of each figure.

B. Describe the strategies you used to find the areas.

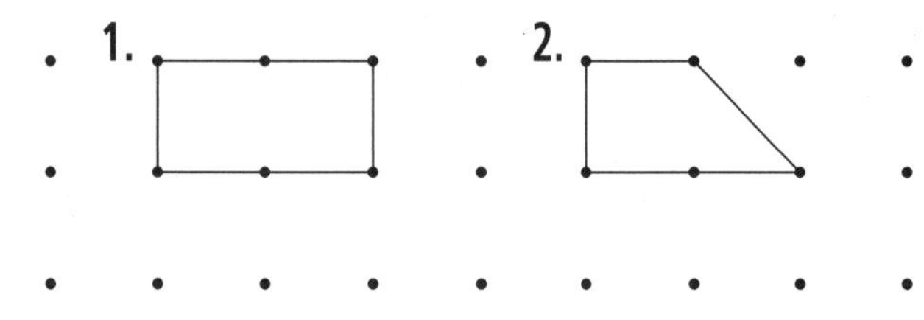

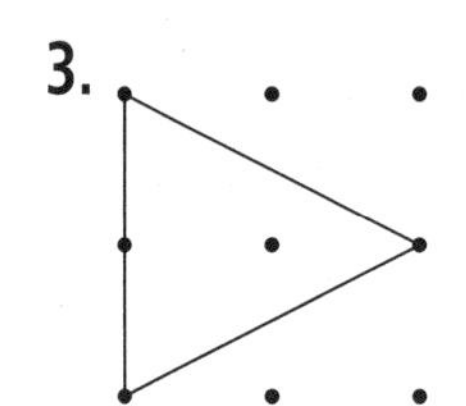

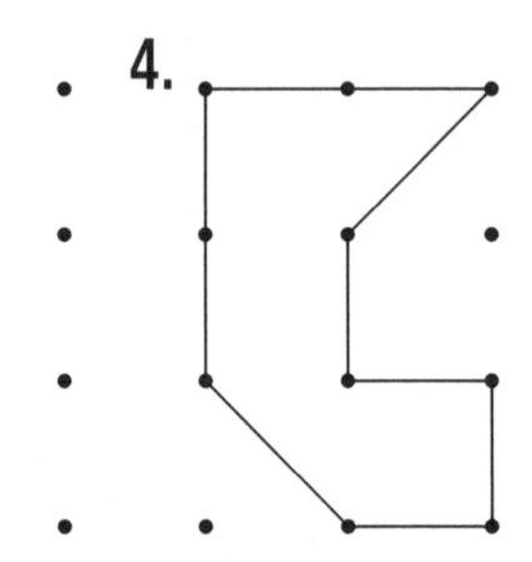

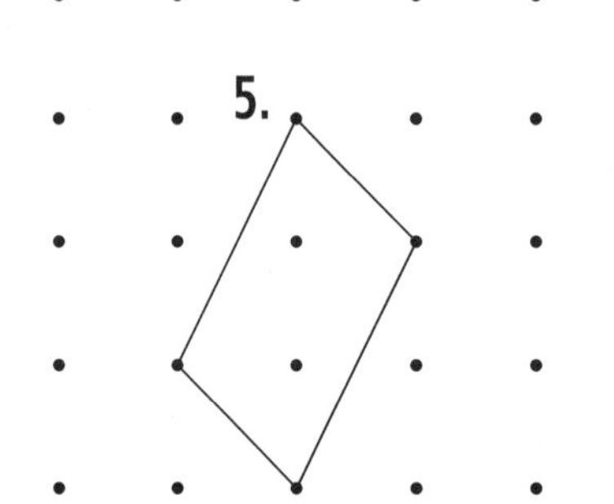

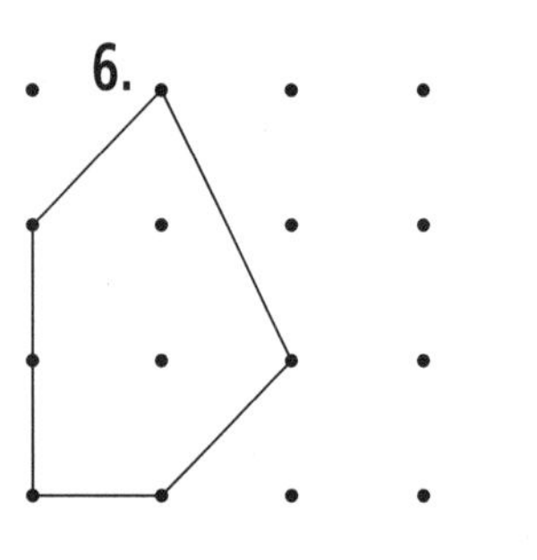

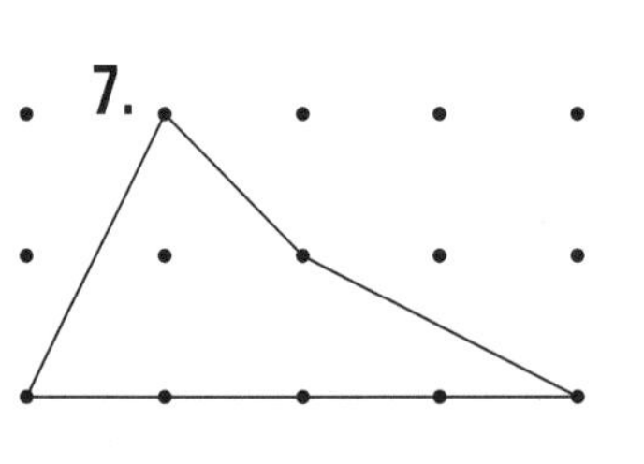

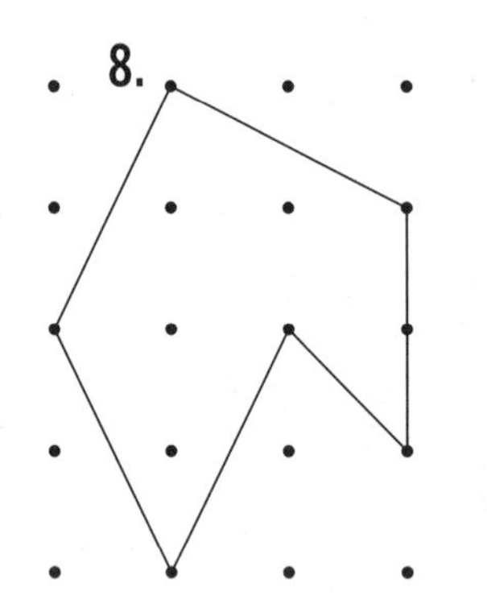

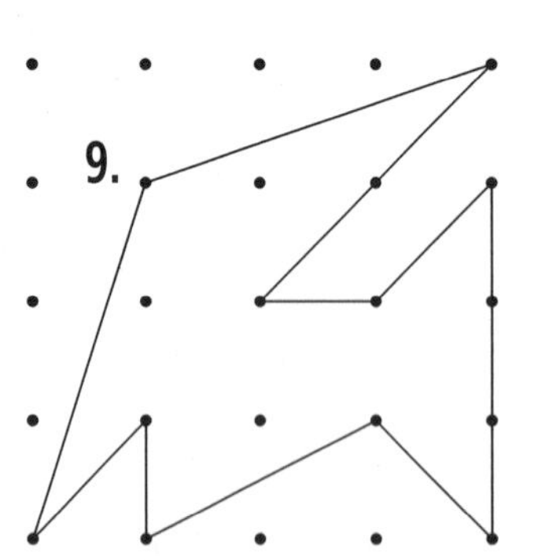

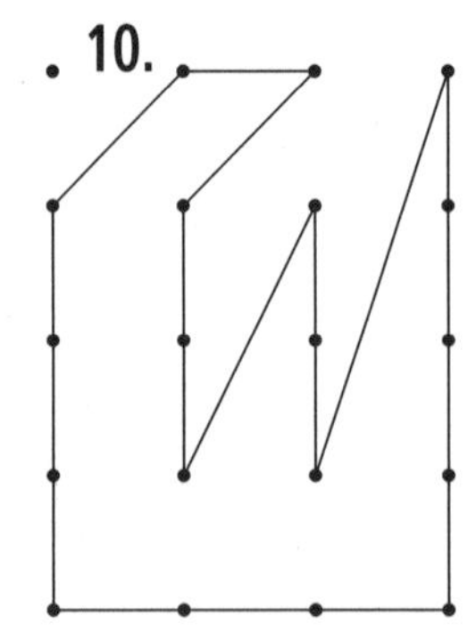

Transparency 2.2

On the 5-dot-by-5-dot grids on Labsheet 2.2, draw squares of various sizes by connecting dots. Try to draw squares with as many different areas as possible. Label each square with its area.

On the 5-dot-by-5-dot grids on Labsheet 2.3, draw line segments of various lengths by connecting dots. Try to draw segments with as many different lengths as possible. Use the method described in your math book to find the length of each segment. To find some of the lengths, you will need to draw squares that extend beyond the 5-dot-by-5-dot grids. Label each segment with its length. Use the $\sqrt{\ }$ symbol to express lengths that are not whole numbers.

Transparency 3.1

A. For each row, draw a right triangle with the given leg lengths on dot paper. Then, draw a square on each side of the triangle.

Length of leg 1	Length of leg 2	Area of square on leg 1	Area of square on leg 2	Area of square on hypotenuse
1	1	1	1	2
1	2			
2	2			
1	3			
2	3			
3	3			
3	4			

B. For each triangle, find the areas of the squares on the legs and on the hypotenuse. Record your results.

C. Look for a pattern in the relationship among the areas of the three squares. Use the pattern you discover to make a conjecture about the relationship among the areas.

D. Draw a right triangle with side lengths that are different from those in the table. Use your triangle to test your conjecture from part C.

A. Cut out the puzzle pieces from Labsheet 3.2. Examine a triangular piece and the three square pieces. How do the side lengths of the squares compare to side lengths of the triangle?

B. Arrange the 11 puzzle pieces to fit exactly into the two puzzle frames. Use four triangles in each frame.

C. Carefully study the arrangements in the two frames. What conclusion can you draw about the relationship among the areas of the three square puzzle pieces?

D. What does the conclusion you reached in part C mean in terms of the side lengths of the triangles?

A. 1. Draw a line segment between points *A* and *B*. Draw a right triangle with segment *AB* as its hypotenuse.

2. Find the lengths of the legs of the triangle.

3. Use the Pythagorean Theorem to find the length of the hypotenuse of the triangle.

B. Use the method described in part A to find the distance between points *C* and *D*.

C. Use the method described in part A to find the distance between points *E* and *F*.

A. 1. Do the whole-number lengths 3, 4, and 5 satisfy the relationship $a^2 + b^2 = c^2$?

2. Form a triangle using string or straws cut to these lengths.

3. Is the triangle you formed a right triangle?

4. Repeat parts 1–3 with the lengths 5, 12, and 13.

5. Make a conjecture about triangles whose side lengths satisfy the relationship $a^2 + b^2 = c^2$.

B. 1. Form a triangle with side lengths a, b, and c that do not satisfy the relationship $a^2 + b^2 = c^2$.

2. Is the triangle a right triangle?

3. Repeat parts 1 and 2 with a different triangle.

4. Make a conjecture about triangles whose side lengths do not satisfy the relationship $a^2 + b^2 = c^2$.

Horace Hanson is the catcher for the Humbolt Bees baseball team. Sneaky Sally Smith, the star of the Canfield Cats, is on first base. Sally is known for stealing bases, so Horace is keeping a sharp eye on her.

The pitcher throws a fastball, and the batter swings and misses. Horace catches the pitch. Out of the corner of his eye, he sees Sally take off for second base.

How far must Horace throw the baseball to get Sally out at second base? Explain how you found your answer.

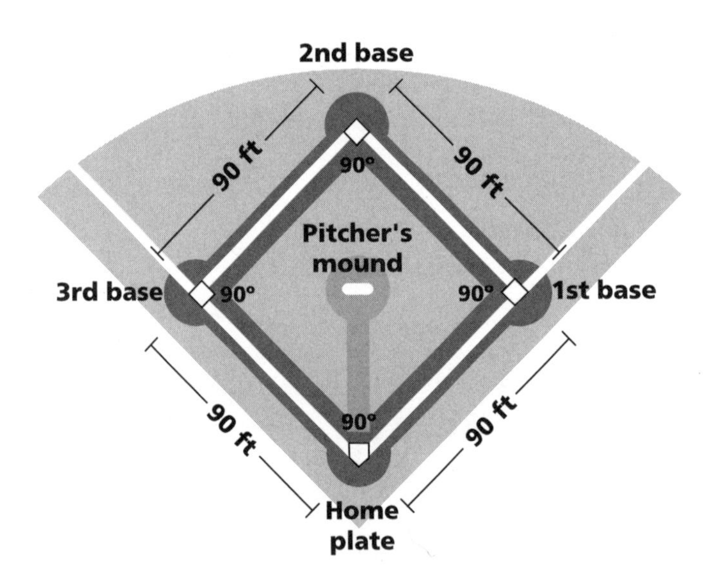

A. How does triangle *ABP* compare with triangle *ACP*?

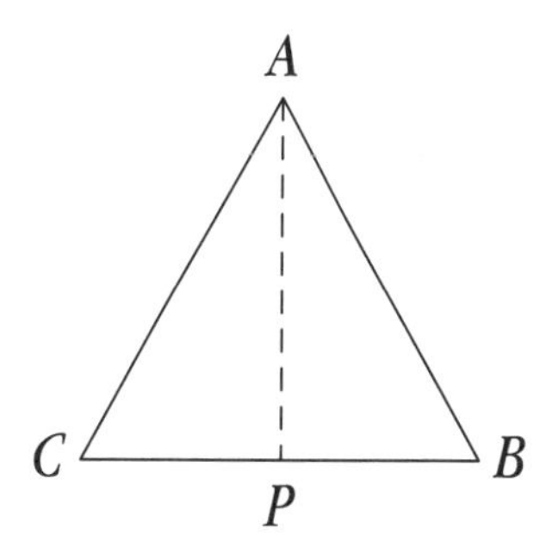

B. Find the measure of each angle in triangle *ABP*. Explain how you found each measure.

C. Find the length of each side of triangle *ABP*. Explain how you found each length.

D. Find a pair of perpendicular line segments in the drawing above.

E. What relationships do you observe among the side lengths of triangle *ABP*? Are these relationships also true for triangle *ACP*? Explain.

In the diagram below, some lengths and angle measures are given. Use this information and what you have learned in this unit to help you find the perimeter of triangle *ABC*. Explain your work.

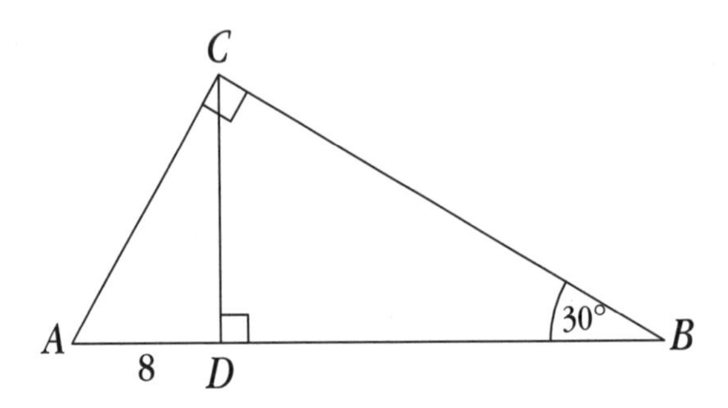

A. Use the Pythagorean Theorem to find the length of each hypotenuse in the Wheel of Theodorus. Label each hypotenuse with its length. Use the $\sqrt{\ }$ symbol to express lengths that are not whole numbers.

B. Cut out the ruler from Labsheet 5.1. Measure each hypotenuse on the Wheel of Theodorus, and label the point on the ruler that represents its length. For example, the first hypotenuse length would be marked like this:

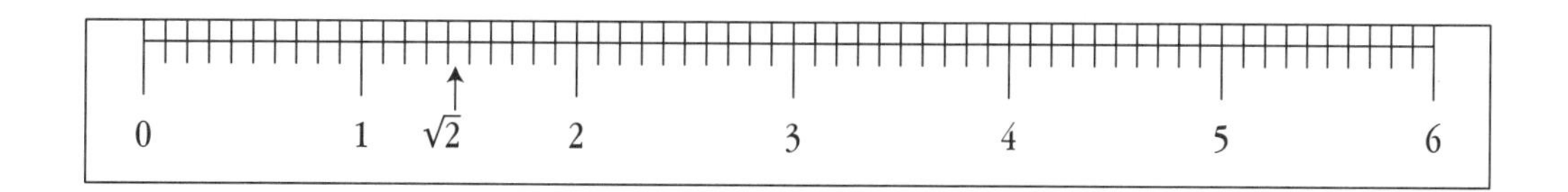

C. For each hypotenuse length that is not a whole number, give the two consecutive whole numbers between which the length is located. For example, $\sqrt{2}$ is between 1 and 2.

D. Use your completed ruler to find a decimal number that is slightly less than each hypotenuse length and a decimal number that is slightly greater than each hypotenuse length. Try to be accurate to the tenths place.

Transparency 5.2

Problem 5.2

Write each fraction as a decimal, and tell whether the decimal is terminating or repeating. If the decimal is repeating, tell which digits repeat.

A. $\frac{2}{5}$

B. $\frac{3}{8}$

C. $\frac{5}{6}$

D. $\frac{35}{10}$

E. $\frac{8}{99}$

A. Copy the table, and write each fraction as a decimal.

Fraction	Decimal	Fraction	Decimal
$\frac{1}{9}$		$\frac{5}{9}$	
$\frac{2}{9}$		$\frac{6}{9}$	
$\frac{3}{9}$		$\frac{7}{9}$	
$\frac{4}{9}$		$\frac{8}{9}$	

B. Describe the pattern you see in your table.

C. Use the pattern to write a decimal representation for each fraction. Use your calculator to check your answers.

1. $\frac{9}{9}$ **2.** $\frac{10}{9}$ **3.** $\frac{15}{9}$

D. What fraction is equivalent to each decimal? Hint: The number 1.222 . . . can be written as 1 + 0.222 . . . .

1. 1.2222 . . . **2.** 2.7777 . . .

Caitlin is playing a video game in which she directs a character named Oskar through a series of obstacles. At this point in the game, Oskar is trapped in the center of an immense forest filled with trees planted in rows.

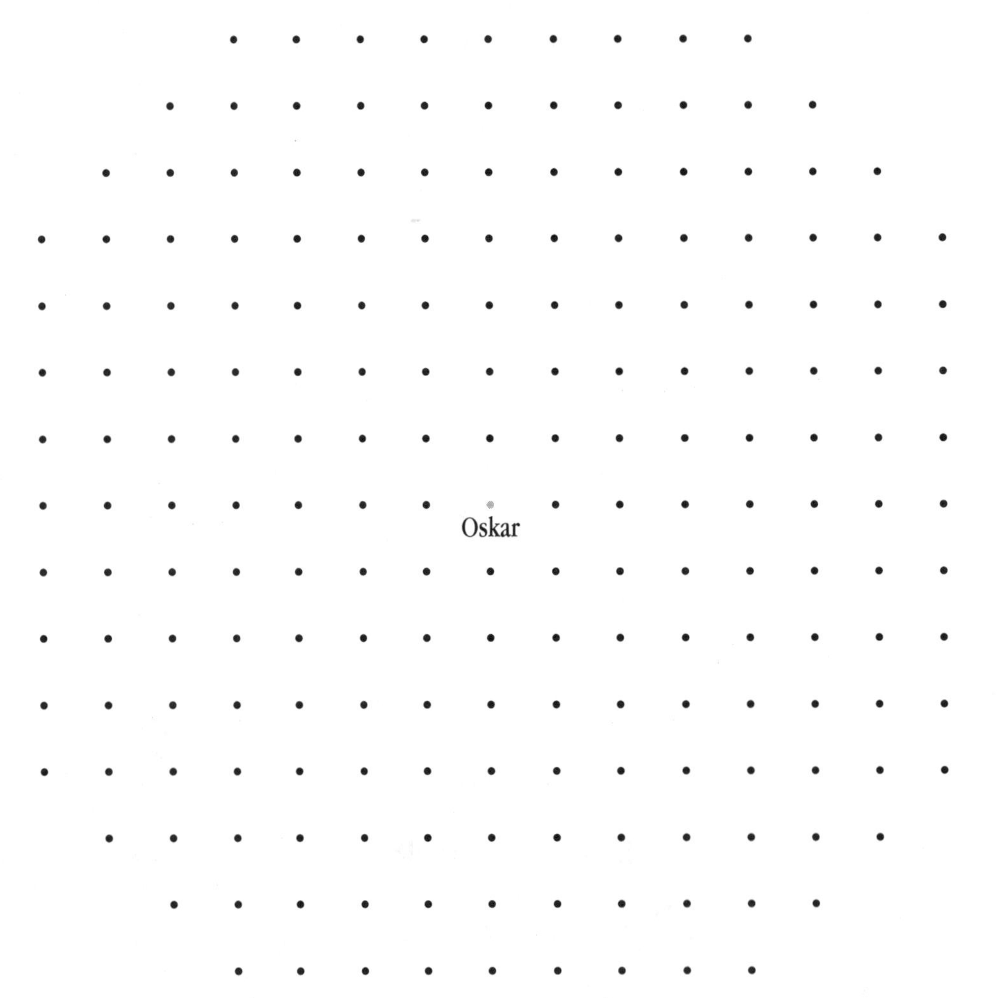

The diagram shows part of the forest. Some of the laser trees have been labeled with letters, and x- and y-axes have been added. Oskar's location is labeled with an O.

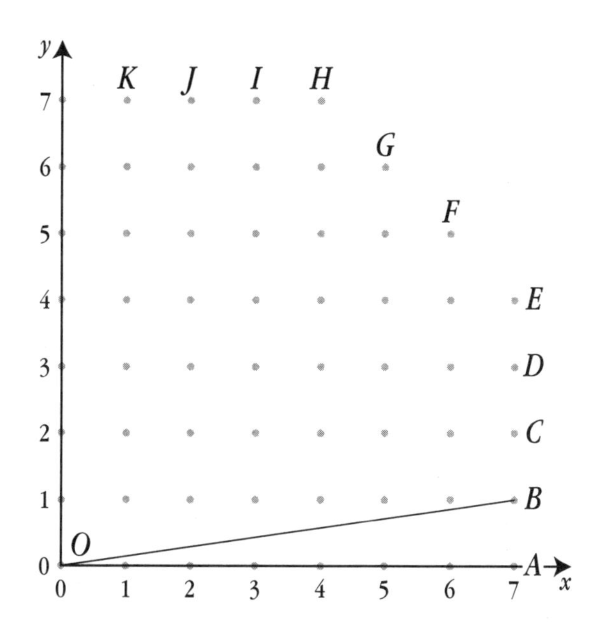

If Oskar had enough power, he could use a laser shield to walk right through the trees. Give the slope of the straight-line path he could follow to get from point O to each of the labeled trees.

The laser forest extends beyond the screen, but Caitlin is not sure how far. She wants to be sure Oskar won't hit a tree anywhere along his escape path.

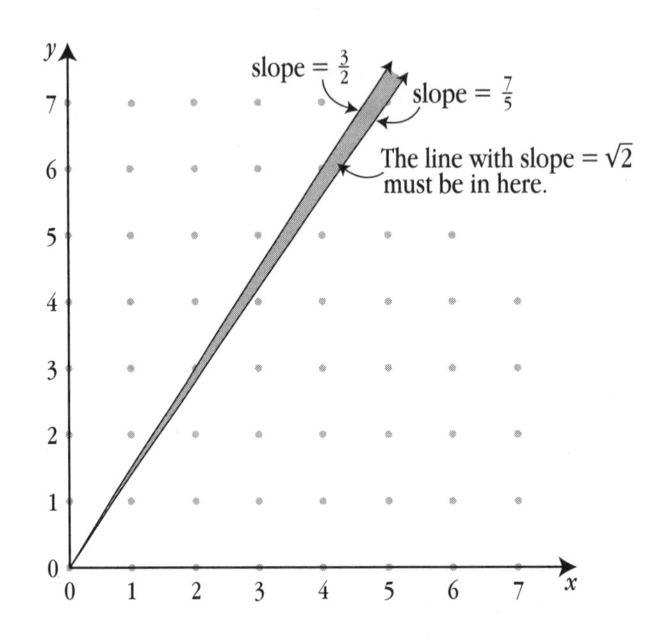

Study the drawing. On Labsheet 6.2, draw a straight-line path that Oskar could follow to get out of the forest. Give the slope of the path you find. Explain why you think your path will work.

Dear Family,

The next unit in your child's course of study in mathematics class this year is *Looking for Pythagoras.* This unit focuses on one of the oldest and most important relationships in all of mathematics, the Pythagorean Theorem. This is the relationship that says that in a right triangle, the sum of the squares of the lengths of the two legs is equal to the square of the length of the longest side, called the hypotenuse. Symbolically, this relationship is $a^2 + b^2 = c^2$, where a and b are the lengths of the legs and c is the length of the hypotenuse.

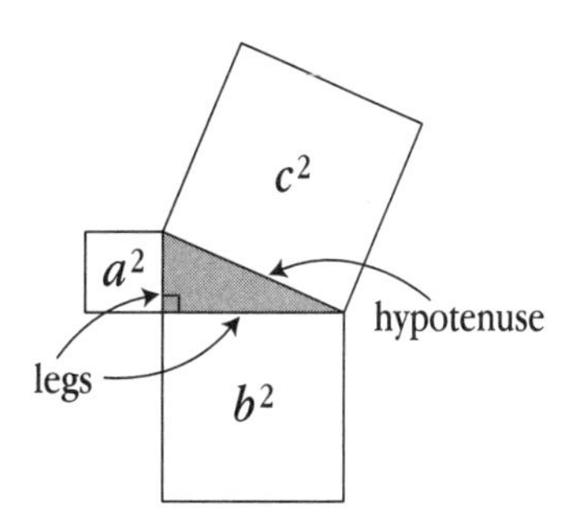

Students use this relationship to compute distances between points on maps and coordinate grids. They apply it to finding the lengths of line segments and other distances. For example, they use it to find out how far a helicopter travels from one point to another and how far a catcher must throw a baseball to get a runner out at second base.

Here are some strategies for helping your child during this unit:

- Ask for an explanation of the ideas presented in the text about finding distances. Help your child to find some examples of right triangles at home or in your community and to apply the Pythagorean Theorem to find the length of one side of a right triangle when the other two are known or can be measured.
- Discuss with your child how the Pythagorean Theorem is applied by people in some careers, such as carpenters, architects, and pilots.
- Encourage your child's efforts in completing all homework assignments. Look over the homework, making sure that all questions are answered and that explanations are clear.

As always, if you have any questions or concerns about this unit or your child's progress, please feel free to call. We want to be sure that this year's mathematics experiences are enjoyable and promote a firm understanding of mathematics.

Sincerely,

Carta para la Familia

Estimada familia,

La próxima unidad del curso de matemáticas de su niño o niña para este año se llama *Looking for Pythagoras* (Busquemos a Pitágoras). Esta unidad está dedicada a una de las relaciones más antiguas y más importantes de las matemáticas, el Teorema de Pitágoras. El Teorema dice que en un triángulo rectángulo, la suma de los cuadrados de los catetos es igual al cuadrado del cateto más largo o hipotenusa. Simbólicamente, esta relación es $a^2 + b^2 = c^2$, donde a y b son las longitudes de los catetos y c es la longitud de la hipotenusa.

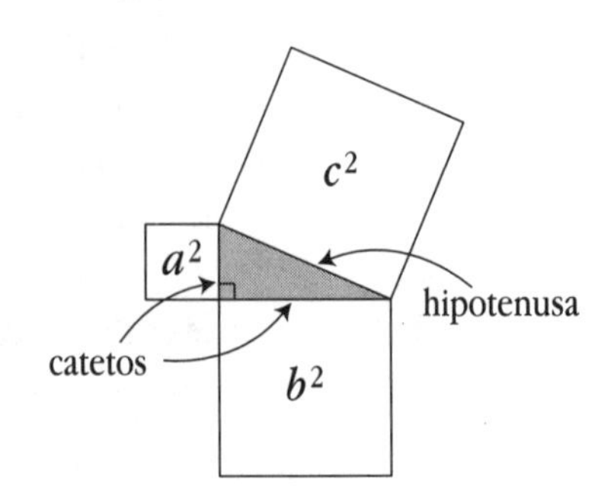

Los estudiantes usan esta relación para computar distancias entre puntos en mapas y cuadrículas de coordenadas. La aplican para hallar longitudes de segmentos lineales y otras distancias. Por ejemplo, para averiguar la distancia recorrida por un helicóptero entre un punto y otro o para saber cuán lejos un apañador debe tirar la pelota de baseball para lograr que el corredor quede afuera en la segunda base.

A continuación les proponemos algunas estrategias para ayudar a su niño o niña durante el desarrollo de la unidad:

- Pídales una explicación de las ideas presentadas en el texto sobre cómo averiguar distancias. Ayúdelos a encontrar ejemplos de triángulos rectángulos en casa o en su barrio y a aplicar el Teorema de Pitágoras para conocer las longitudes de uno de los lados del triángulo rectángulo cuando los otros dos son conocidos o se pueden medir.

- Discuta con su niño o niña cómo gente de diferentes carreras usa el Teorema de Pitágoras, por ejemplo carpinteros, arquitectos o pilotos.

- Aliente a su niño o niña a completar las tareas asignadas para el hogar. Revise sus tareas, asegurándose de que todas las preguntas han sido contestadas y de que las explicaciones son claras.

Como ya es habitual, si Ud. tiene alguna pregunta o preocupación acerca de esta unidad o acerca del progreso de su niño o niña, por favor no dude en visitarnos. Queremos estar seguros de que las experiencias dentro del campo de las matemáticas sean agradables para los niños y niñas este año, y también queremos promover una buena comprensión de la materia.

Atentamente,

Isometric

Dot Paper

Centimeter Grid Paper

Additional Practice

Investigation 1

Use these problems for additional practice after Investigation 1.

Refer to the map on page 8 to answer 1–3.

1. Which landmarks are 5 blocks apart by car?

2. The taxi stand is 5 blocks by car from the hospital and 5 blocks by car from the police station. Give the coordinates of the taxi stand.

3. The airport is halfway between City Hall and the hospital by helicopter. Give the coordinates of the airport.

4. **a.** Draw a square with vertices (0, 1), (1, 0), (0, $^-$1), and ($^-$1, 0). What is the area of this square in small triangles?

   b. Draw a square around the square you made in part a with two of the vertices at (1, 1) and ($^-$1, 1). What are the other two vertices? What is the area of this square in small triangles?

   c. Draw the square of the next size. One of its vertices is (0, $^-$2). What are the other three vertices? What is the area of this square in small triangles?

   small triangle

   d. What are the four vertices of the square of the next size? What is its area in small triangles?

   e. What do you notice about the areas of the squares as the squares get larger?

5. Use the Venn diagram below to answer parts a–f. The Venn diagram shows a way to think about the classification of quadrilaterals.

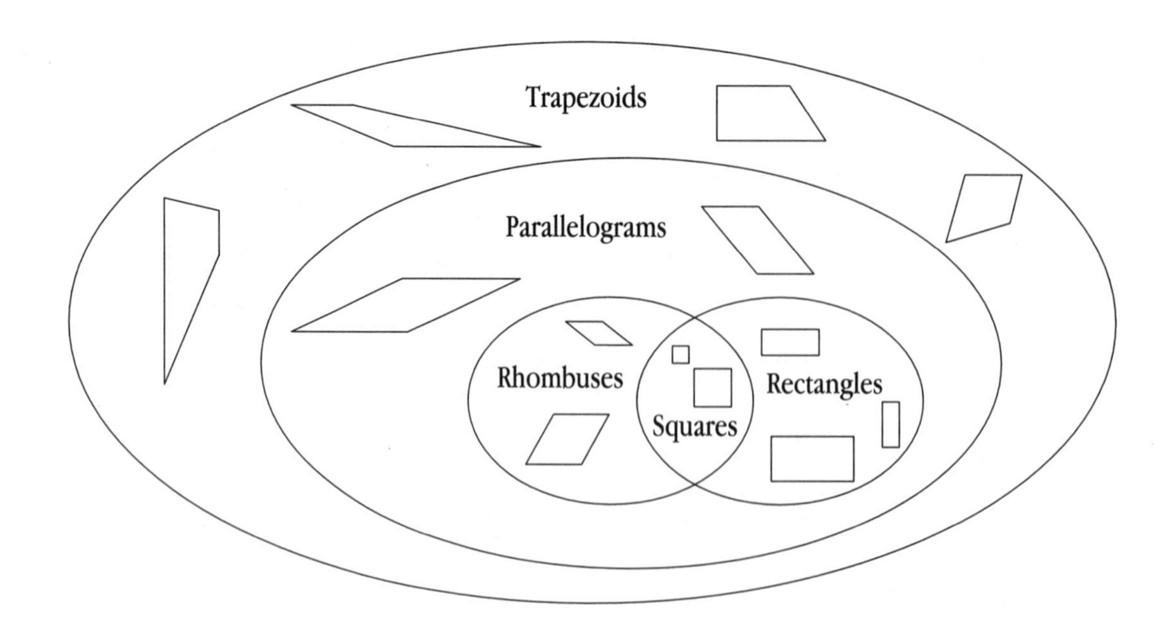

a. Which quadrilaterals have only one pair of parallel sides?

b. Which quadrilaterals always have four right angles?

c. Which quadrilaterals have two pairs of parallel sides?

d. *All squares are rectangles.* Is this statement true or false? If it is false, show an example that disproves it.

e. *Some rectangles are rhombuses.* Is this statement true or false? If it is false, show an example that disproves it.

f. *All trapezoids are rhombuses.* Is this statement true or false? If it is false, show an example that disproves it.

6. In parts a–d, draw the figure described. Recall that the sum of the angles in any quadrilateral is 360°.

   a. Draw a quadrilateral in which at least two angles measure 90°.

   b. Draw a rhombus in which at least one angle measures 45°.

   c. Draw a trapezoid in which the angles measure either 40° or 140°.

   d. Draw a parallelogram in which the angles measure either 40° or 140°.

In 7–12, use the given lengths to find the area of the figure. Show your calculations. Think about which formulas you can use as part of your reasoning.

7.

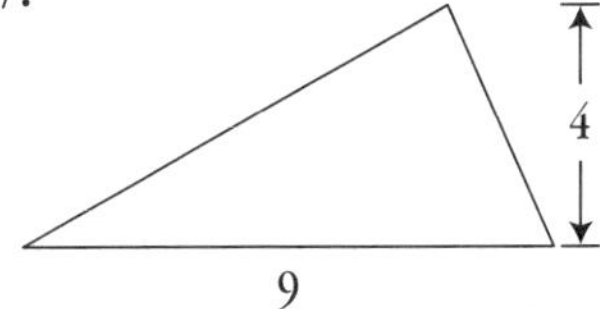

8.

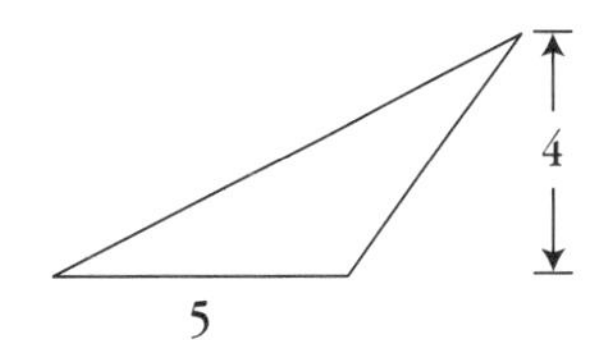

9.

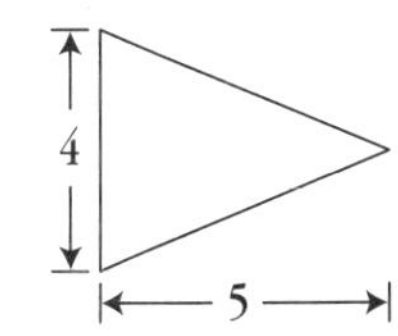

10.

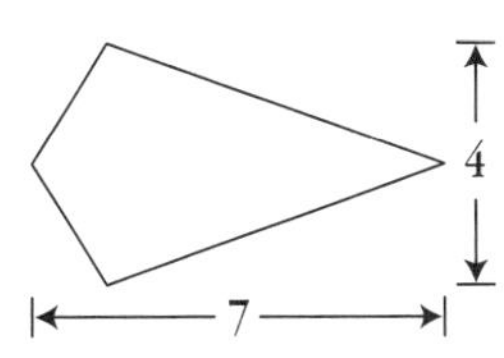

11.

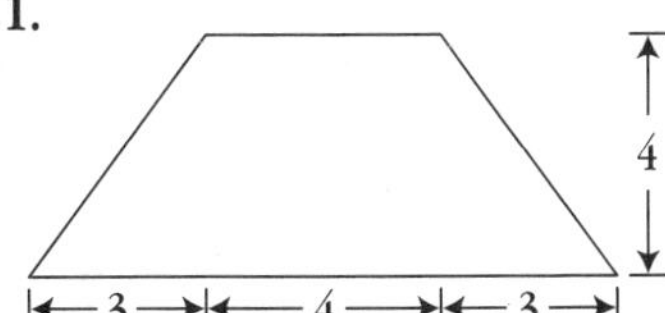

12.

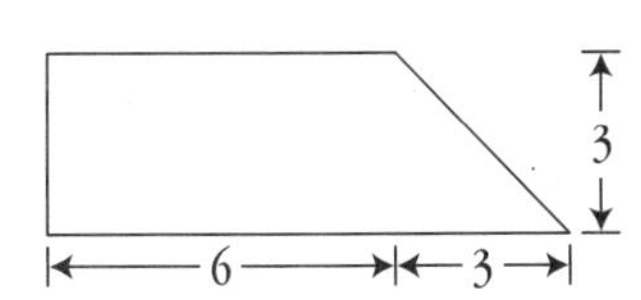

Investigation 2

Use these problems for additional practice after Investigation 2.

In 1–4, find the area of the figure. Describe the method you use.

1.

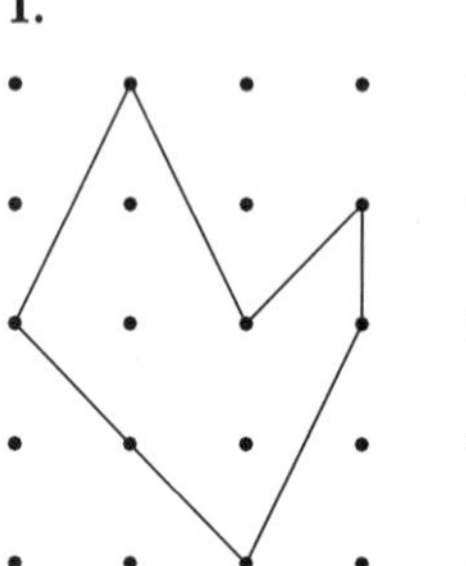

2.

3.

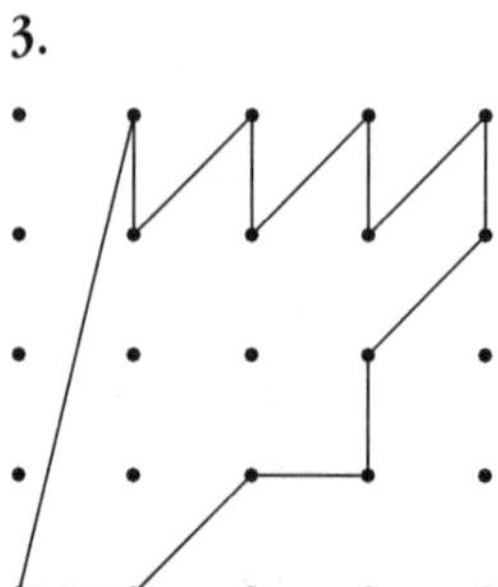

4.

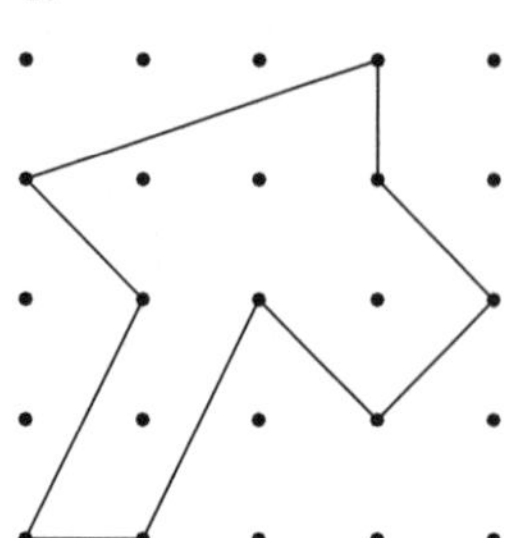

In Problem 2.3, you found the lengths of line segments drawn on 5-dot-by-5-dot grids. Some of those lengths were written as square roots, such as $\sqrt{2}$. When you enter $\sqrt{2}$ in your calculator, the result is a decimal with a value of approximately 1.4. In 5–10, find the approximate value for the given length to the nearest tenth.

5. $\sqrt{5}$

6. $\sqrt{13}$

7. $\sqrt{20}$

8. $\sqrt{17}$

9. $\sqrt{2} + \sqrt{5}$

10. $\sqrt{8} + 6 + \sqrt{10}$

11. Is $\sqrt{8} + \sqrt{10}$ the same as $\sqrt{8 + 10}$? Prove your answer in two ways:

a. Use your calculator to help give a numerical argument.

b. Use a grid and lengths of line segments to give a geometric argument.

In 12–14, find the perimeter of the figure. Express the perimeter in two ways: as the sum of a whole number and square roots, and as a single value after using decimal approximations to the nearest tenth for the square roots. An example is done for you.

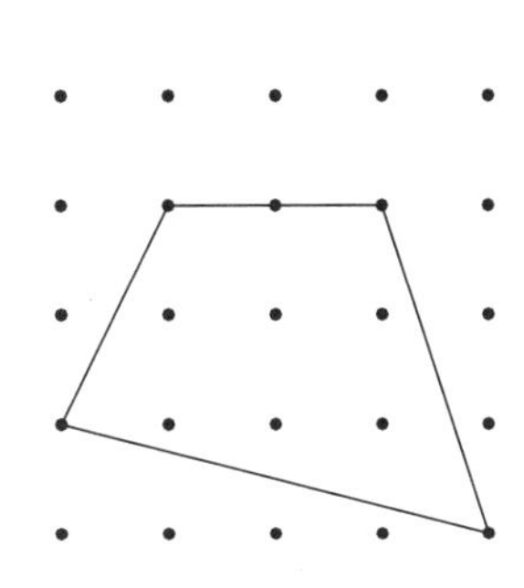

The perimeter of this figure is $2 + \sqrt{10} + \sqrt{17} + \sqrt{5}$

$\approx 2 + 3.2 + 4.1 + 2.2$

≈ 11.5

12.

13.

14.

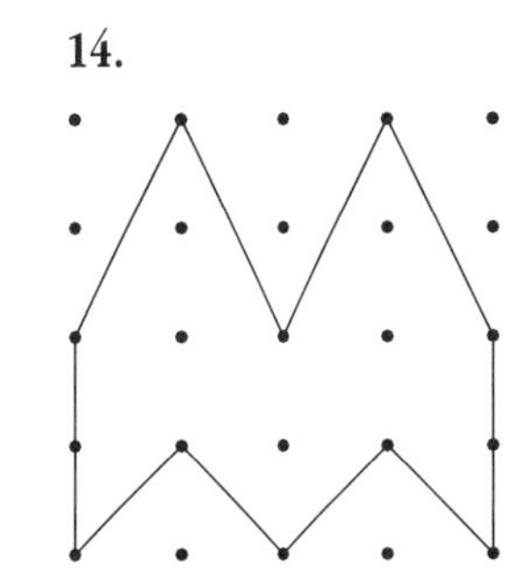

Use this information for 15–17: Triangles may be classified by angle measures or side lengths.

Classification by angles	Classification by side lengths
Acute triangle: all angles < 90°	*Equilateral triangle:* all sides same length
Right triangle: one angle = 90°	*Isosceles triangle:* two sides same length
Obtuse triangle: one angle > 90°	*Scalene triangle:* all sides different lengths

15. Recall that the sum of the measures of the angles in any triangle is 180°. There is one triangle listed above in which all three angles have the same measure. Which triangle is this? How did you decide?

16. Can you draw an isosceles right triangle on a 5-dot-by-5-dot grid? If so, draw one. If not, explain why not.

17. Can you draw a scalene triangle on a 5-dot-by-5-dot grid? If so, draw one. If not, explain why not.

In 18–21, copy the triangle. Use what you know about side lengths and angle measures to find the missing angle and side measures.

18.

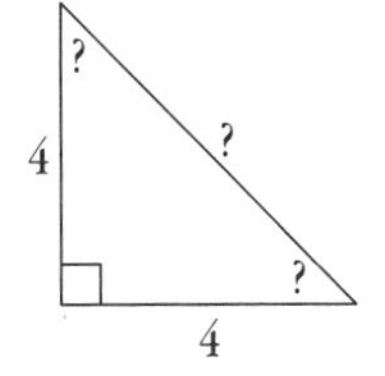

19.

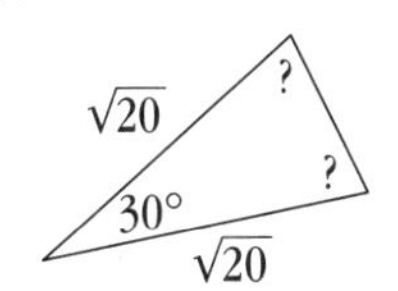

20.

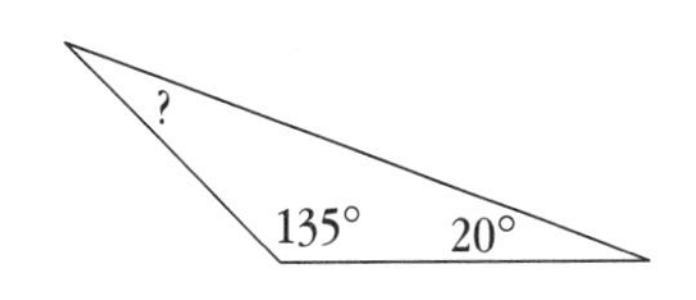

21.

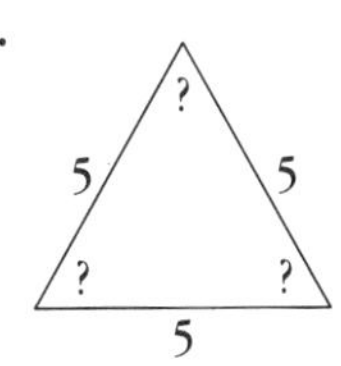

Investigation 3

Use these problems for additional practice after Investigation 3.

1. Consider the right triangles shown below.

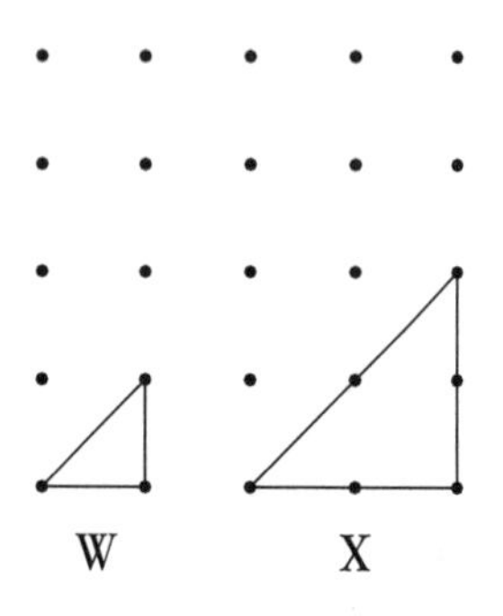

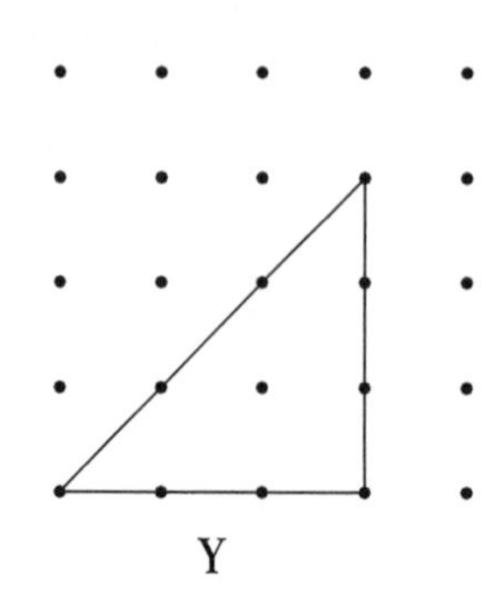

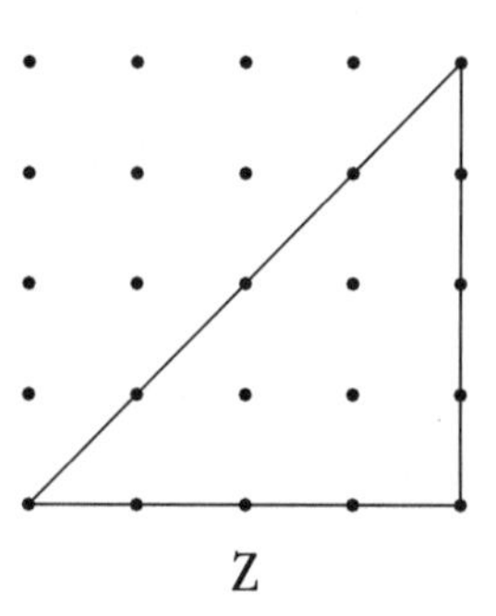

   a. Find the length of the hypotenuse of each triangle.

   b. How are the hypotenuse lengths in figures X, Y, and Z related to the hypotenuse length in figure W?

2. Draw a right triangle with a hypotenuse length of $\sqrt{5}$.

3. Draw a right triangle with a hypotenuse length of $2\sqrt{5}$.

4. Draw a right triangle with a hypotenuse length of $3\sqrt{5}$.

5. Give the coordinates of two points on a coordinate grid that are $\sqrt{10}$ apart.

6. Give the coordinates of two points that are $\sqrt{13}$ apart.

7. Give the coordinates of two points that are $\sqrt{32}$ apart.

8. Give the coordinates of two points that are $7\sqrt{2}$ apart.

9. Give the coordinates of a point on a coordinate grid that is a distance of $\sqrt{5}$ from point (1, 3).

10. Give the coordinates of a point that is a distance of $\sqrt{17}$ from point $(0, ^-5)$.

11. Give the coordinates of a point that is a distance of $2\sqrt{5}$ from point $(^-10, ^-2)$.

12. Give the coordinates of a point that is a distance of $3\sqrt{5}$ from point $(8, ^-2)$.

13. What is the length of the line that connects points (0, 0) and (4, 2)?

14. What is the length of the line that connects points (0, 0) and (2, 4)?

15. What is the length of the line that connects points $(^-2, 0)$ and (0, 2)?

16. What is the length of the line that connects points $(0, ^-3)$ and (3, 3)?

In 17–19, find the perimeter of the figure to the nearest tenth.

17.

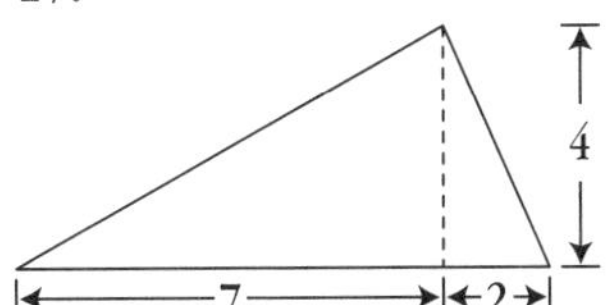

18.

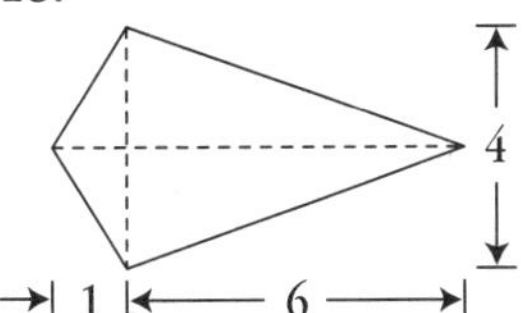

19.

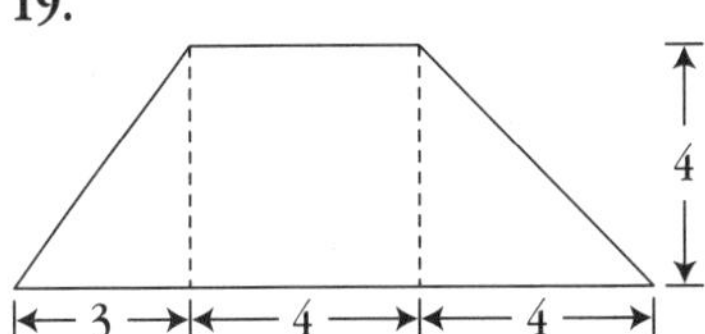

In 20–23, use the map on page 8 to find the distance by helicopter between the two landmarks, and explain how you found the distance.

20. the greenhouse and the police station

21. the police station and the art museum

22. the greenhouse and City Hall

23. City Hall and the animal shelter

In 24–26, find the perimeter of the right triangle. Express the perimeter in two ways: as the sum of a whole number and square roots, and as a single value after using decimal approximations to the nearest tenth for the square roots. An example is done for you.

The perimeter of this figure is $4 + \sqrt{10} + \sqrt{18}$

$\approx 2 + 3.2 + 4.2$

≈ 9.4

24.

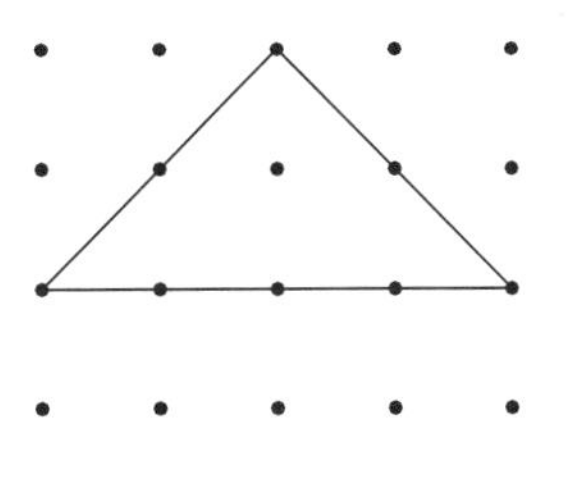

25.

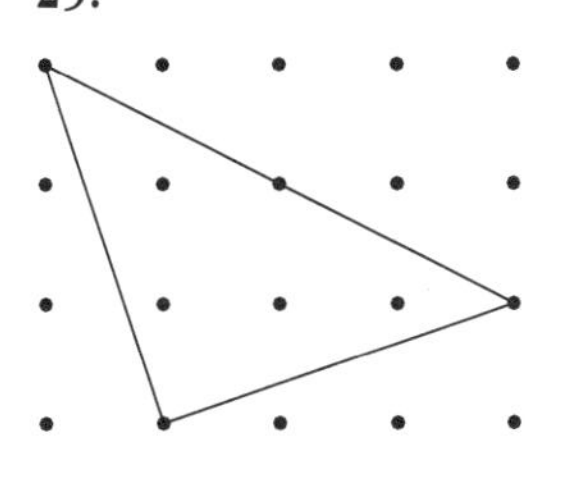

26.

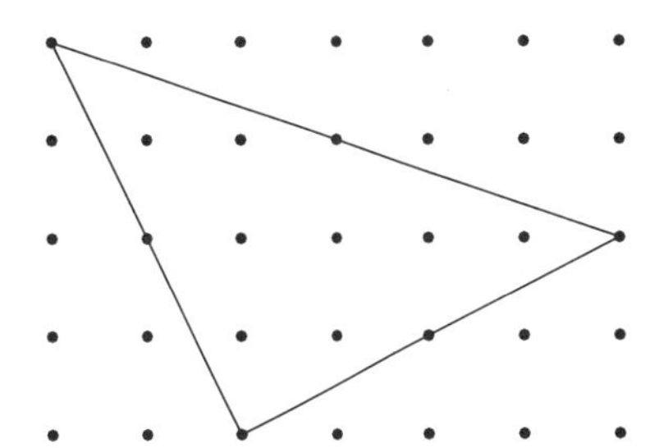

Additional Practice

In 27–30, find the area of the figure. Describe the method you use.

27.

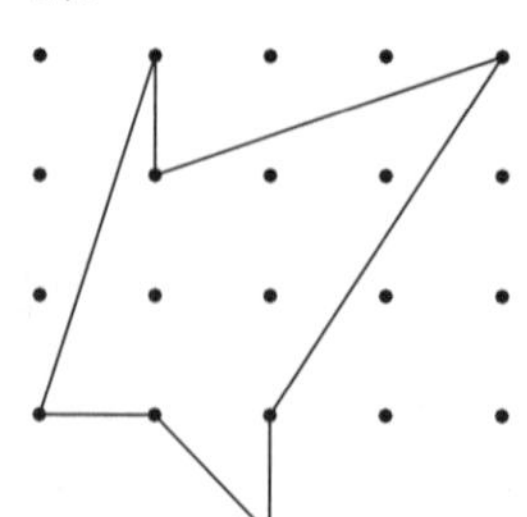

28.

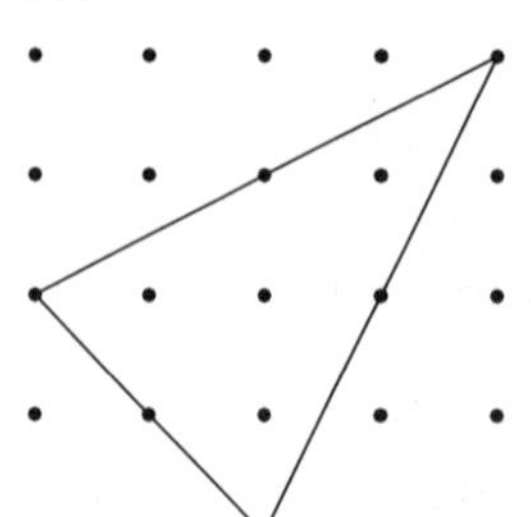

29.

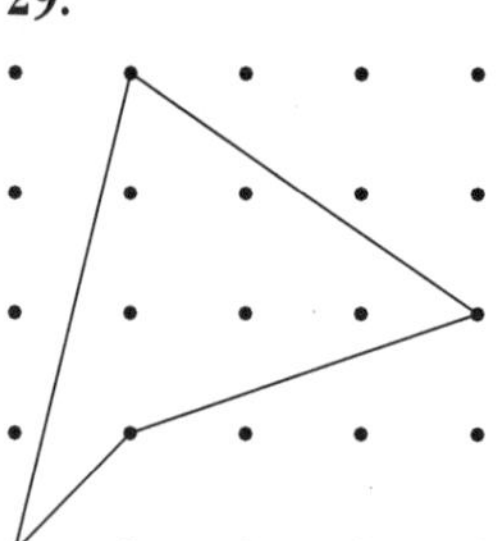

30.

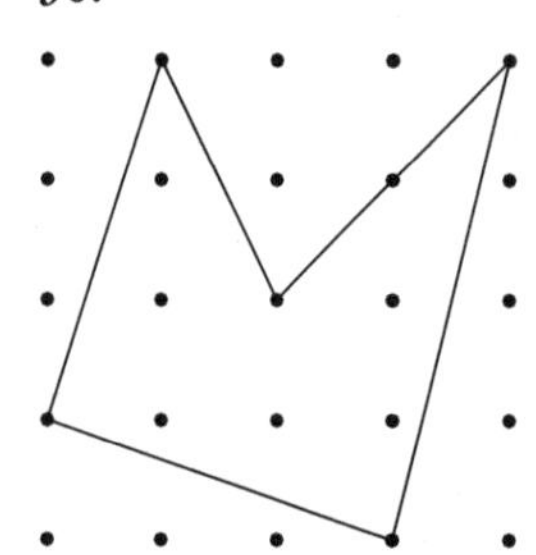

In 31–33, find the perimeter of the figure. Express the perimeter in two ways: as the sum of a whole number and square roots, and as a single value after using decimal approximations to the nearest tenth for the square roots.

31.

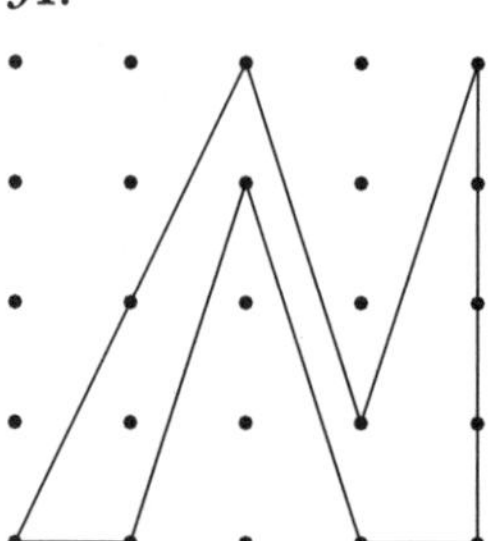

32.

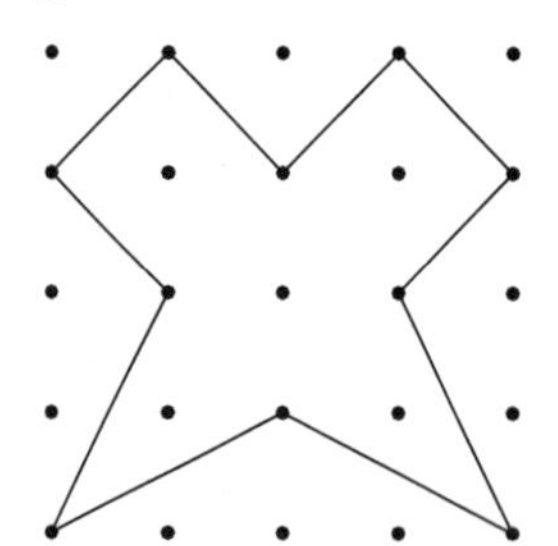

33.

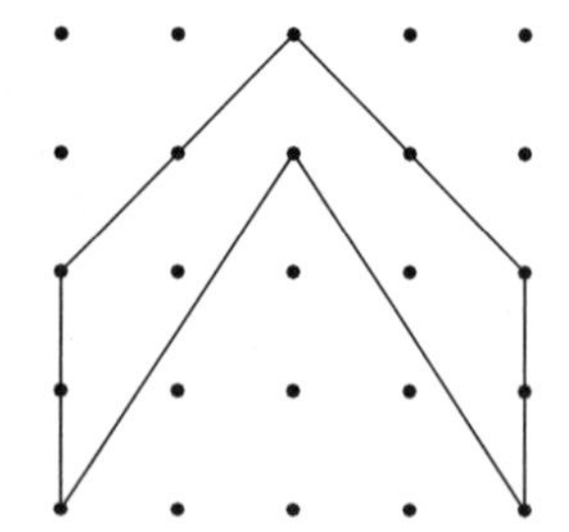

Investigation 4

Use these problems for additional practice after Investigation 4.

In 1–4, find the length of AB to the nearest hundredth. Show how you find the length.

1.

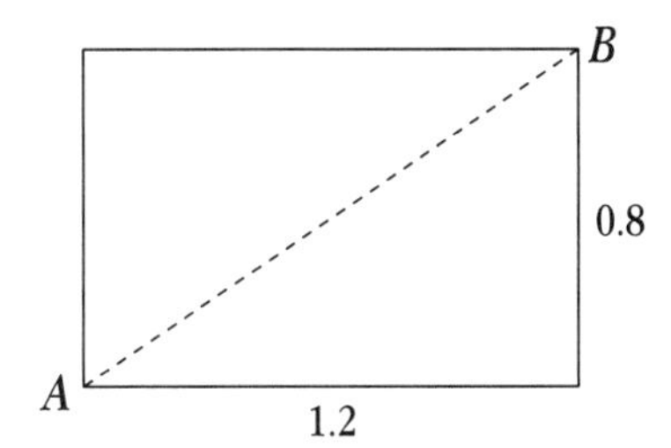

2. This is a regular pentagon.

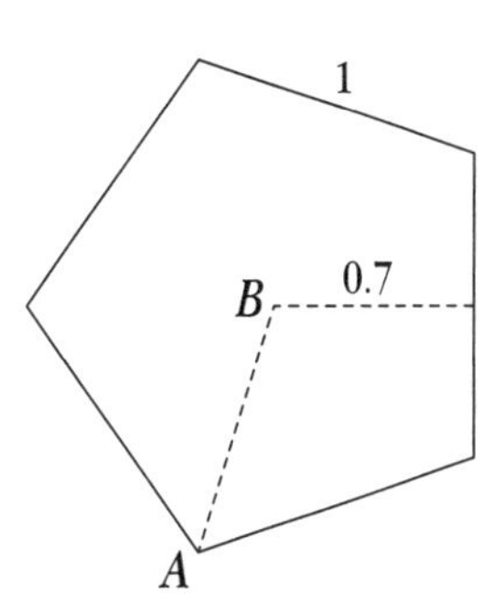

3.

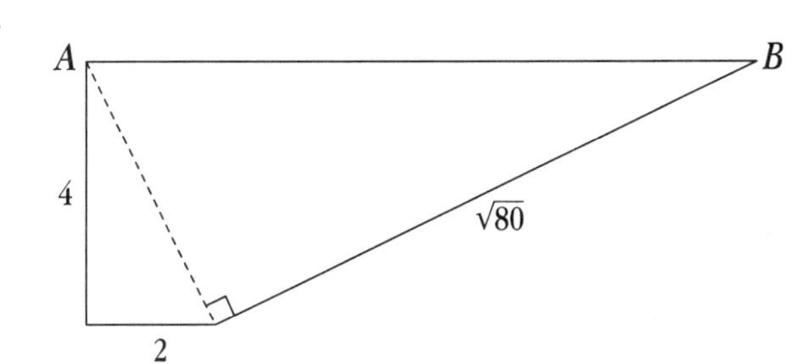

4. This is a regular hexagon.

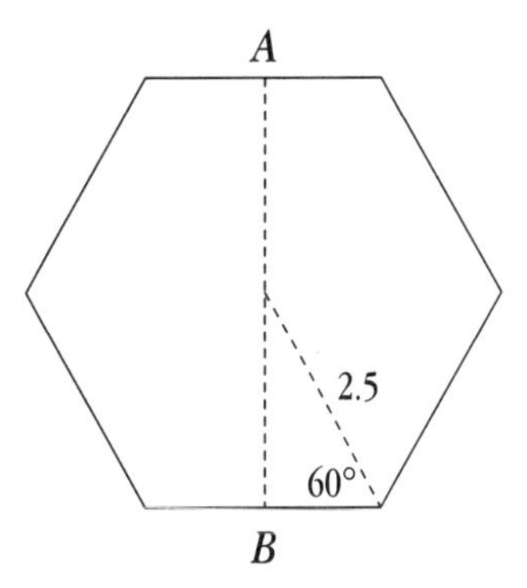

In 5–8, find the perimeter of the figure to the nearest tenth.

5.

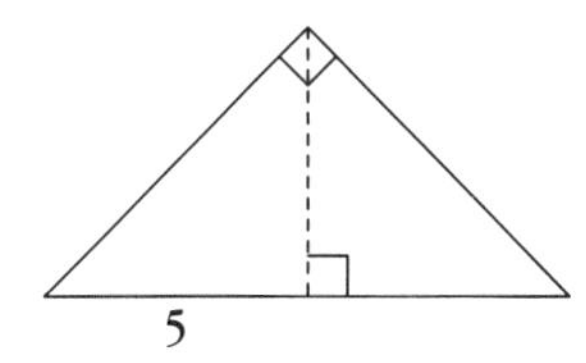

6.

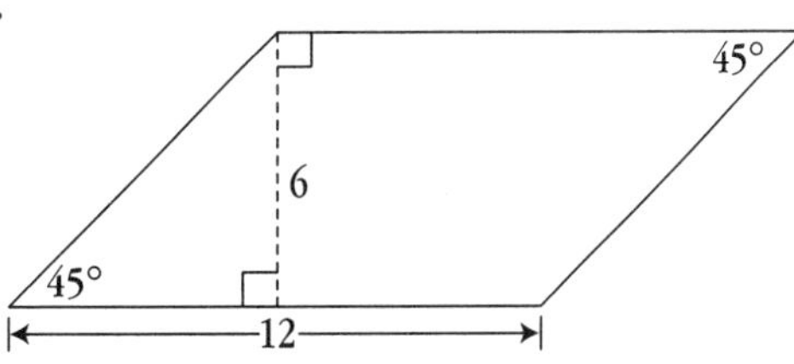

7.

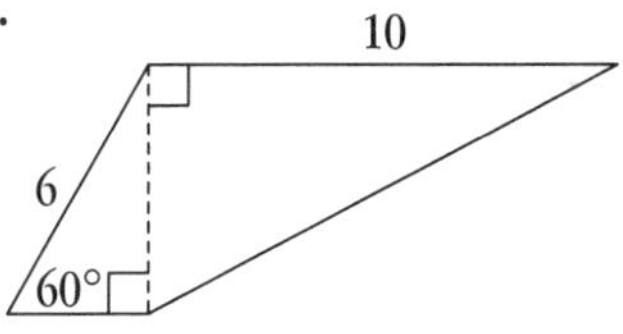

8.

12
6
4

9. **a.** Find the areas of figures W and X. Describe the method you use.

   b. Draw two different figures Y and Z, each with an area of $7\frac{1}{2}$ square units. Be clever!

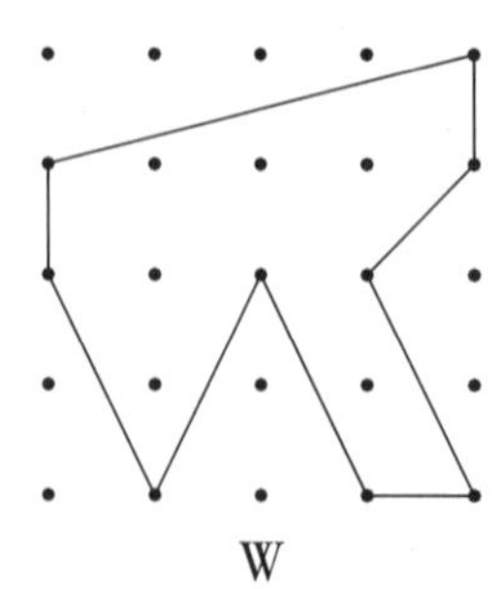

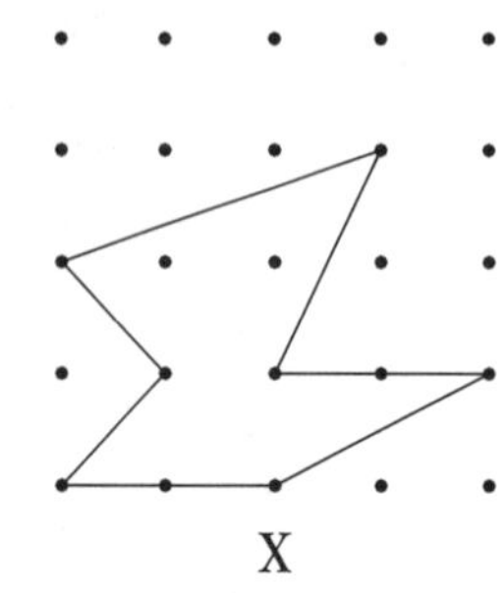

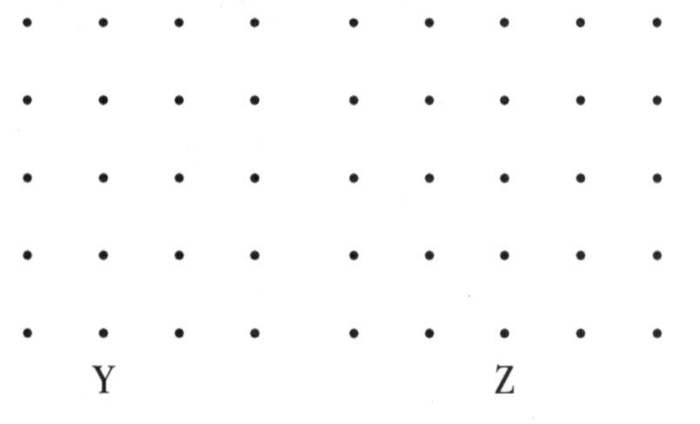

An isosceles right triangle has two legs that are the same length. In 10 and 11, sketch the triangle described, and label the three side lengths.

10. Two of the sides in this isosceles right triangle measure $\sqrt{18}$ and 3.

11. Two of the sides in this isosceles right triangle measure $\sqrt{52}$ and $\sqrt{26}$.

In 12–17, a pair of lengths are given. What third length could be used with the other two lengths to make a right triangle? Try to solve each problem two ways: (1) let the missing value be the length of one of the legs of the triangle and (2) let the missing value be the length of the hypotenuse of the triangle. Sketch each triangle you find, and label the side lengths.

12. 9, 15, and ?

13. 3, $\sqrt{45}$, and ?

14. $\sqrt{50}$, 5, and ?

15. $\sqrt{18}$, 3, and ?

16. 8, $\sqrt{18}$, and ?

17. $\sqrt{52}$, $\sqrt{26}$, and ?

Investigation 5

Use these problems for additional practice after Investigation 5.

1. a. What are the decimal representations of $\frac{1}{111}$, $\frac{2}{111}$, $\frac{3}{111}$, and $\frac{4}{111}$?

   b. Find other fractions with decimal representations the same as those of $\frac{1}{111}$, $\frac{2}{111}$, $\frac{3}{111}$, and $\frac{4}{111}$.

2. a. What are the decimal representations of $\frac{1}{33}$, $\frac{2}{33}$, $\frac{3}{33}$, and $\frac{4}{33}$?

   b. Find other fractions with decimal representations the same as those of $\frac{1}{33}$, $\frac{2}{33}$, $\frac{3}{33}$, and $\frac{4}{33}$.

3. a. What are the decimal representations of $\frac{1}{333}$, $\frac{2}{333}$, $\frac{3}{333}$, and $\frac{4}{333}$?

   b. Find other fractions with decimal representations the same as those of $\frac{1}{333}$, $\frac{2}{333}$, $\frac{3}{333}$, and $\frac{4}{333}$.

In 4–9, give a decimal representation of the fraction.

4. $\frac{43}{9}$ 5. $\frac{63}{99}$ 6. $\frac{1000}{333}$

7. $\frac{29}{11}$ 8. $\frac{290}{333}$ 9. $\frac{870}{999}$

In 10–12, give a fraction representation of the decimal.

10. 0.242424 . . . 11. 0.080808 . . . 12. 4.323232 . . .

13. Name three fraction representations of the decimal 0.363636 . . . .

14. In parts a–e, give the two consecutive whole numbers between which the given number is located.

    a. $\sqrt{17}$ b. $\sqrt{83}$ c. $\sqrt{250}$ d. $\sqrt{400}$ e. $\sqrt{1650}$

    f. How can you use your calculator to answer parts a–e?

In 15–18, tell whether the statement is true or false.

15. $7 = \sqrt{49}$ 16. $\sqrt{6.25} = 0.25$

17. $13 = \sqrt{149}$ 18. $\sqrt{777} = 27$

Investigation 6

Use these problems for additional practice after Investigation 6.

1. Find the slope of each line segment on the grid.

   AL *BL* *KL* *LD* *IJ* *JD* *AJ* *JE* *GD* *BD*

G F E

K

H D

J

I L

C

A B

2. Find each figure in the grid above, and match it with one of the names in the list. Names may be used more than once.

   a. figure *HGFD*
   b. figure *HJI*
   c. figure *FKE*
   d. figure *ABI*
   e. figure *ELC*
   f. figure *JKLB*

   Figures
   parallelogram
   scalene triangle (triangle with three different side lengths)
   trapezoid
   right triangle
   rectangle
   square
   isosceles triangle
   rhombus

3. Find the perimeter of each figure named in question 2. Approximate your answer to the nearest hundredth.

Investigation 1

1. the gas station and the art museum; the stadium and City Hall; the cemetery and the animal shelter
2. $(^-3, ^-2)$
3. $(^-3, ^-2)$
4. The four squares are shown at right.

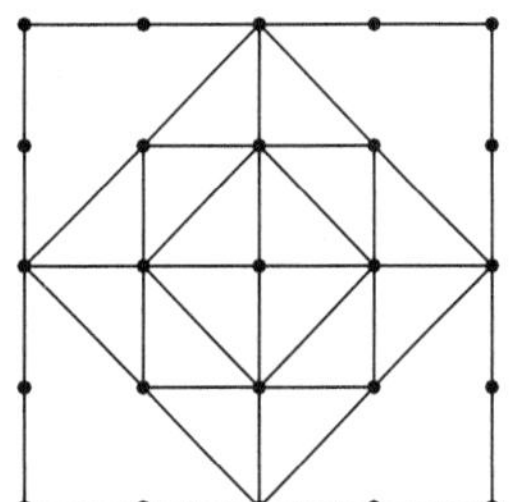

   a. area = 4 small triangles
   b. $(^-1, ^-1)$, $(1, ^-1)$; area = 8 small triangles
   c. $(2, 0)$, $(0, 2)$, $(^-2, 0)$; area = 16 small triangles
   d. $(2, 2)$, $(^-2, 2)$, $(^-2, ^-2)$, $(2, ^-2)$; area = 32 small triangles
   e. The area of each new square is twice the area of the previous square. (Note: The area increases by powers of 2: 2^2, 2^3, 2^4, 2^5.)
5. a. trapezoids
   b. rectangles and squares
   c. parallelograms, rhombuses, rectangles, and squares
   d. true
   e. true; All squares are rectangles and rhombuses.
   f. false; Here is a trapezoid that is not a rhombus:

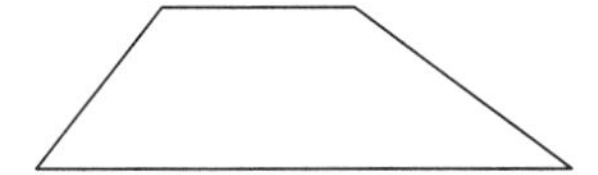

6. a. Possible answers:

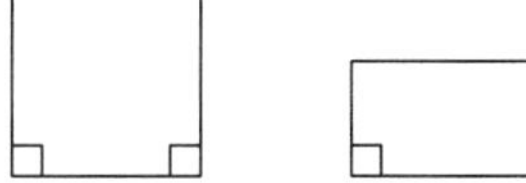

   b. Possible answers:

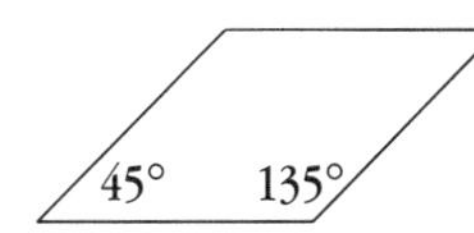

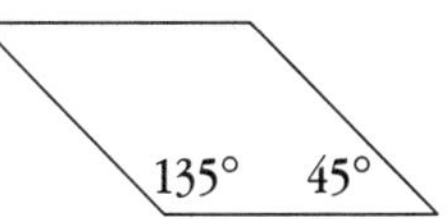

   c. Possible answers:

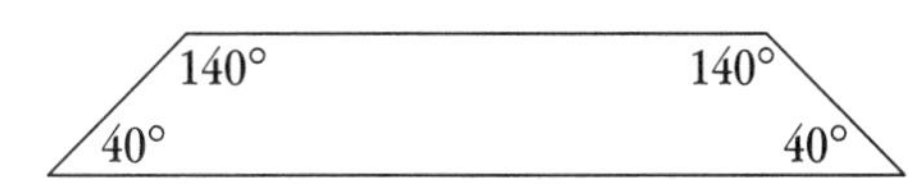

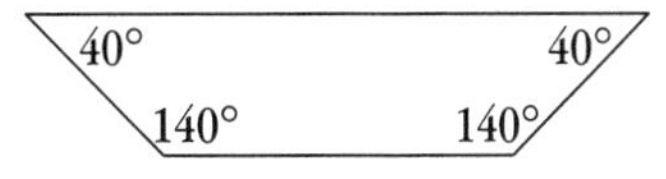

d. Possible answers:

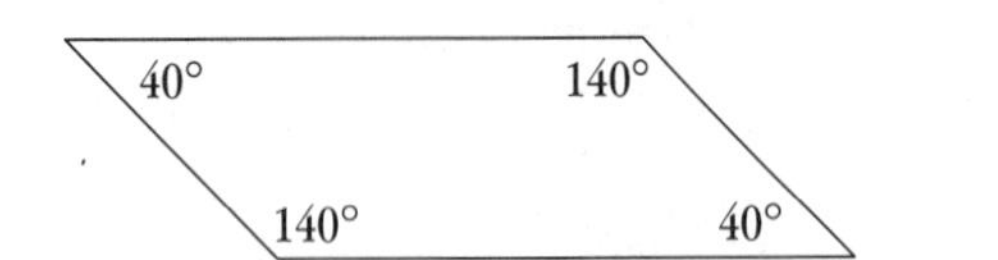

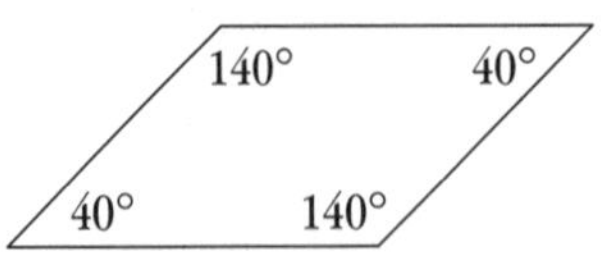

7. The area of this triangle is $\frac{1}{2}bh = \frac{1}{2}(9)(4) = 18$ square units.
8. The area of this triangle is $\frac{1}{2}bh = \frac{1}{2}(5)(4) = 10$ square units.
9. Think of the figure as two triangles, each with base 5 and height 2. Then, area $= 2(\frac{1}{2}bh) = 2(\frac{1}{2})(5)(2) = 10$ square units.
10. Think of the figure as two triangles, each with base 7 and height 2. Then, area $= 2(\frac{1}{2}bh) = 2(\frac{1}{2})(7)(2) = 14$ square units.
11. Think of the figure as two triangles, each with base 3 and height 4, and a square with side length 4. Then, area $= 2(\frac{1}{2}bh) + s^2 = 2(\frac{1}{2})(3)(4) + 4^2 = 12 + 16 = 28$ square units.
12. Think of the figure as a triangle with base 3 and height 3 and a rectangle with side lengths 6 and 3. Then, area $= \frac{1}{2}bh + lw = \frac{1}{2}(3)(3) + 6 \times 3 = 4.5 + 18 = 22.5$ square units.

Investigation 2

1. $5\frac{1}{2}$ square units; Methods will vary.
2. 7 square units; Methods will vary.
3. $8\frac{1}{2}$ square units; Methods will vary.
4. 7.5 square units; Methods will vary.
5. $\sqrt{5} \approx 2.2$
6. $\sqrt{13} \approx 3.6$
7. $\sqrt{20} \approx 4.5$
8. $\sqrt{17} \approx 4.1$
9. $\sqrt{2} + \sqrt{5} \approx 3.7$
10. $\sqrt{8} + 6 + \sqrt{10} \approx 12.0$
11. **a.** Since $\sqrt{8} + \sqrt{10} \approx 6.0$ and $\sqrt{8 + 10} = \sqrt{18} \approx 4.2$, the expressions are not equal.

    b. Drawing segments on dot paper shows that these are not equivalent, as the two shorter segments obviously don't add to the third segment.

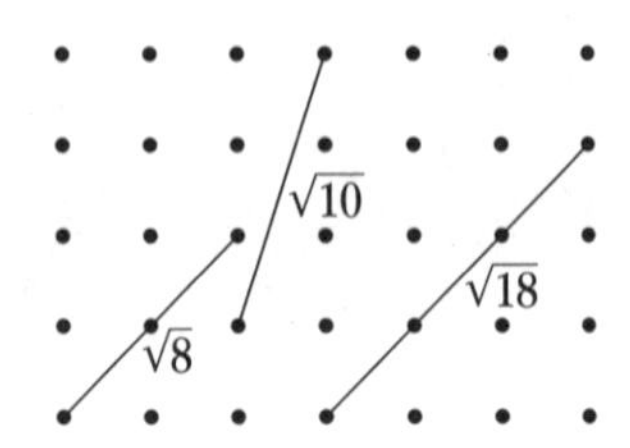

12. $3 + \sqrt{10} + \sqrt{17} + \sqrt{18} \approx 3 + 3.2 + 4.1 + 4.2 = 14.5$

13. $\sqrt{25} + \sqrt{10} + \sqrt{17} \approx 5 + 3.2 + 4.1 = 12.3$

14. $4 + 4\sqrt{5} + 4\sqrt{2} \approx 4 + 4(2.2) + 4(1.4) = 18.4$

15. In an equilateral triangle, each angle is 60°. You can measure the angles to see that each is 60°.

16. yes; Possible triangle:

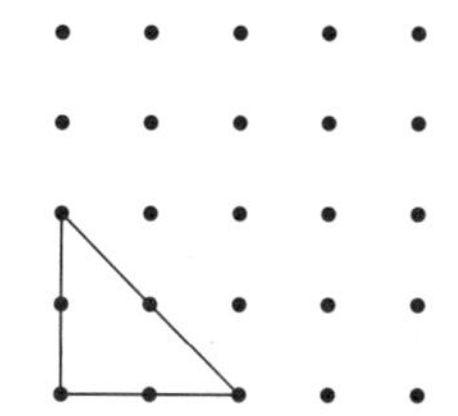

17. yes; Possible triangle:

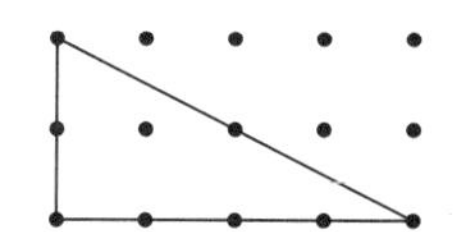

18.

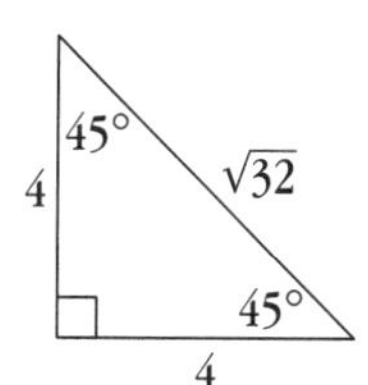

19.

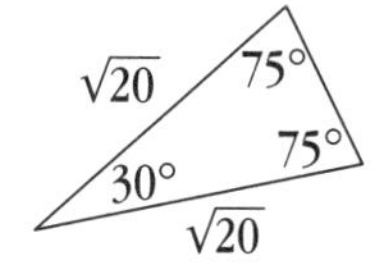

20.

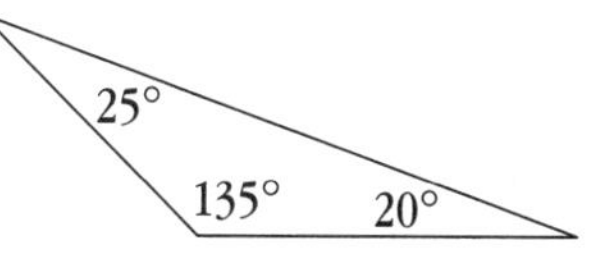

21.

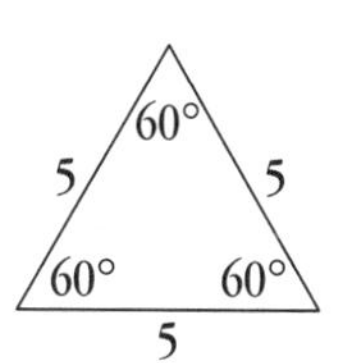

Investigation 3

1. a. triangle W, $\sqrt{2}$; triangle X, $\sqrt{8}$; triangle Y, $\sqrt{18}$; triangle Z = $\sqrt{32}$

   b. The hypotenuse of figure X is 2 times as long, the hypotenuse of figure Y is 3 times as long, and the hypotenuse of figure Z is 4 times as long as the hypotenuse of figure W.

2.

3. The length $2\sqrt{5} = \sqrt{20}$.

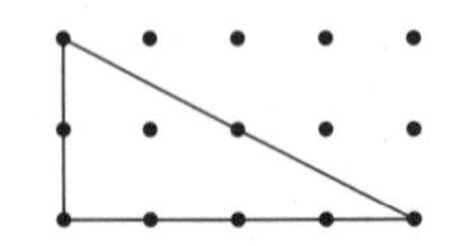

4. The length $3\sqrt{5} = \sqrt{45}$.

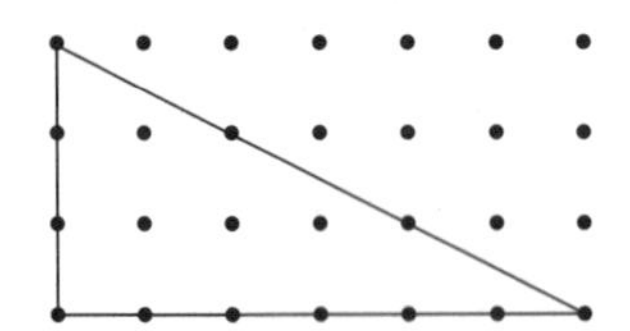

5. Possible answer: (0, 0) and (3, 1)
6. Possible answer: (0, 0) and (3, 2)
7. Possible answer: (0, 0) and (4, 4)
8. Possible answer: (0, 0) and (7, 7)
9. There are eight possible points: (2, 1), (3, 2), (3, 4), (2, 5), (0, 5), (-1, 4), (-1, 2), and (0, 1).
10. There are eight possible points: (-1, -1), (1, -1), (4, -4), (4, -6), (1, -9), (-1, -9), (-4, -6), and (-4, -4).
11. There are eight possible points: (-8, 2), (-12, 2), (-14, 0), (-14, -4), (-12, -6), (-8, -6), (-6, -4), and (-6, 0).
12. There are eight possible points: (2, 1), (5, 4), (11, 4), (14, 1), (14, -5), (11, -8), (5, -8), and (2, -5).
13. $\sqrt{20}$
14. $\sqrt{20}$
15. $\sqrt{8}$
16. $\sqrt{45}$
17. perimeter = $9 + \sqrt{20} + \sqrt{65} \approx 21.5$

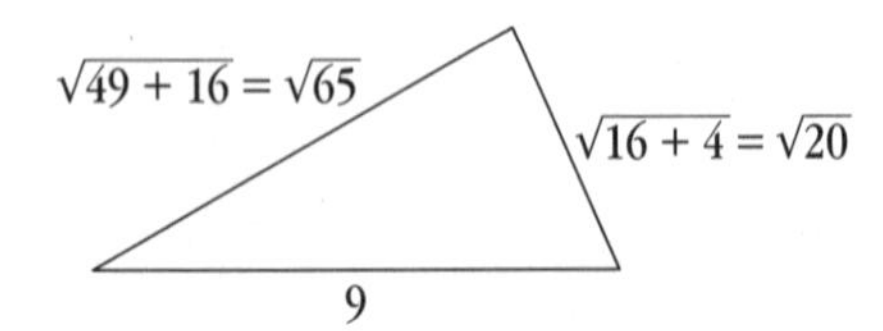

18. perimeter = $2\sqrt{5} + 2\sqrt{40} \approx 17.1$

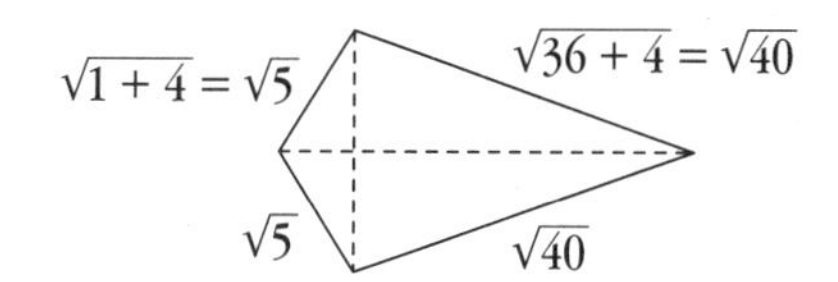

19. perimeter = $\sqrt{32} + \sqrt{25} + 4 + 11 \approx 25.7$

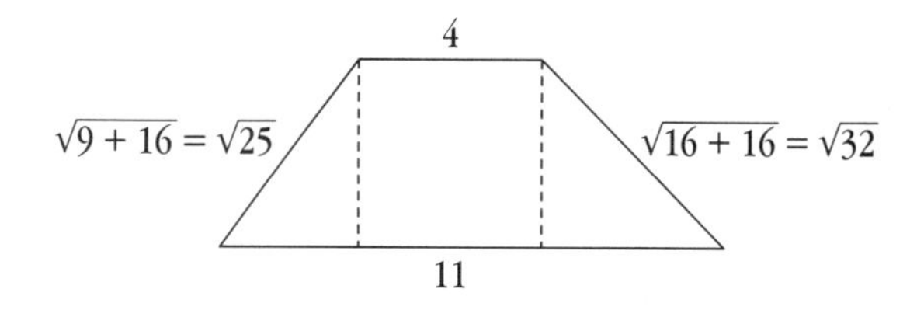

20. $\sqrt{36 + 16} = \sqrt{52} \approx 7.2$ blocks

21. $\sqrt{36 + 25} = \sqrt{61} \approx 7.8$ blocks

22. 6 blocks

23. $\sqrt{4 + 36} = \sqrt{40} \approx 6.3$ blocks

24. $4 + 2\sqrt{8} \approx 4 + 2(2.8) = 9.6$

25. $2\sqrt{10} + \sqrt{20} \approx 2(3.2) + 4.5 = 10.9$

26. $2\sqrt{20} + \sqrt{40} \approx 2(4.5) + 6.3 = 15.3$

27. 6 1/2 square units; Methods will vary.

28. 6 square units; Methods will vary.

29. 6 square units; Methods will vary.

30. 8 square units; Methods will vary.

31. $6 + 4\sqrt{10} + \sqrt{20} \approx 6 + 4(3.2) + 4.5 = 23.3$

32. $6\sqrt{2} + 4\sqrt{5} \approx 6(1.4) + 4(2.2) = 17.2$

33. $4 + 2\sqrt{8} + 2\sqrt{13} \approx 4 + 2(2.8) + 2(3.6) = 16.8$

Investigation 4

Note: Some answers given here include sums or differences under a radical sign; students may express their solutions in different ways.

1. $1.2^2 + 0.8^2 = 2.08$; $AB = \sqrt{2.08} \approx 1.44$

2. $0.5^2 + 0.7^2 = 0.74$; $AB = \sqrt{0.74} \approx 0.86$

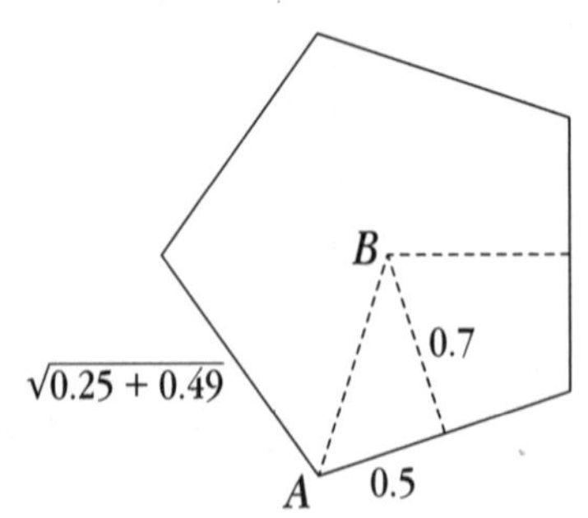

3. $AB = \sqrt{100} = 10$

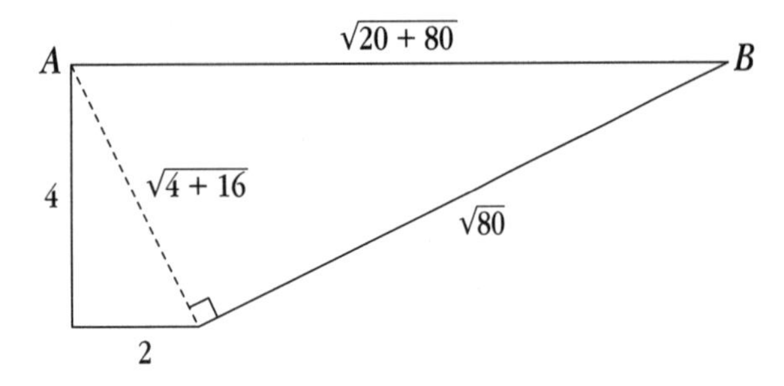

4. $2\sqrt{4.69} \approx 4.33$

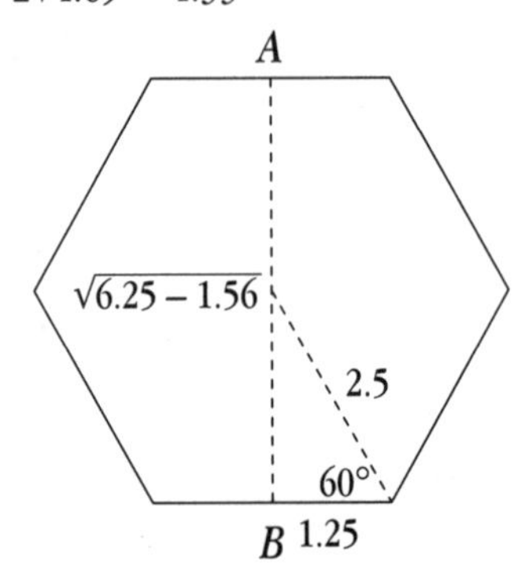

5. perimeter $= 10 + 2\sqrt{50} \approx 24.1$

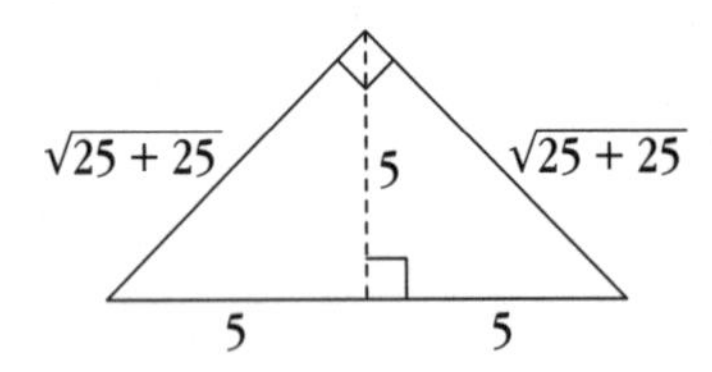

6. perimeter $= 24 + 2\sqrt{72} \approx 41.0$

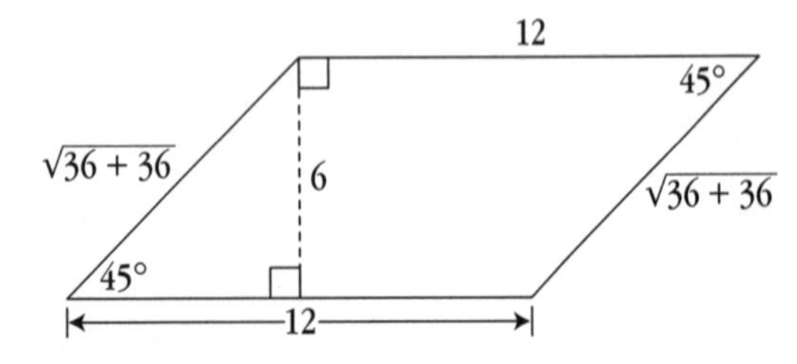

7. perimeter = $19 + \sqrt{127} \approx 30.3$

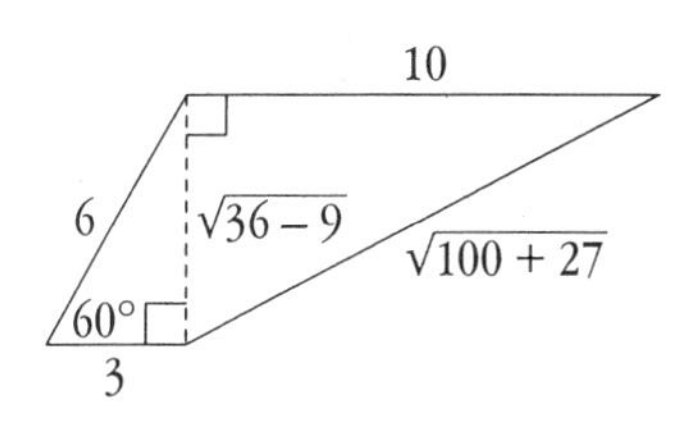

8. perimeter = $16 + \sqrt{108} + \sqrt{52} \approx 33.6$

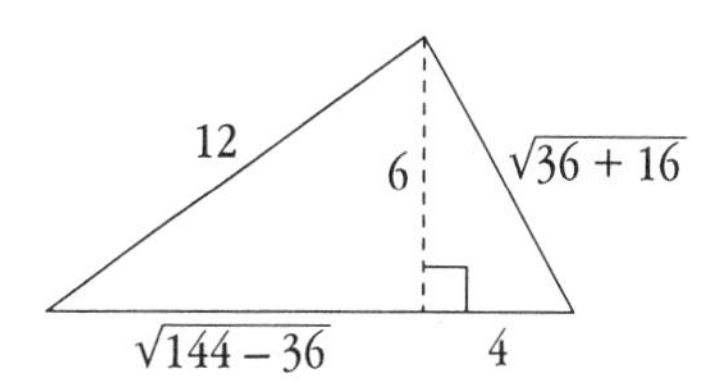

9. a. figure W: $9\frac{1}{2}$ square units; figure X: $5\frac{1}{2}$ square units; Methods will vary.

   b. Answers will vary.

10.

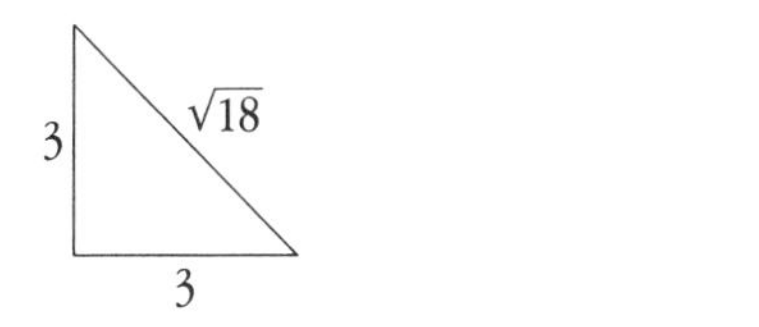

11.

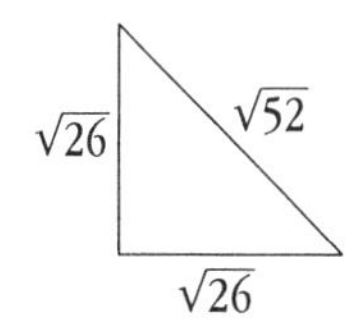

12.

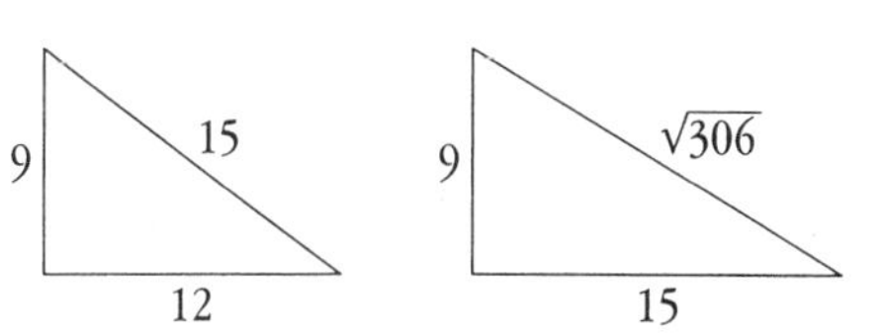

13.

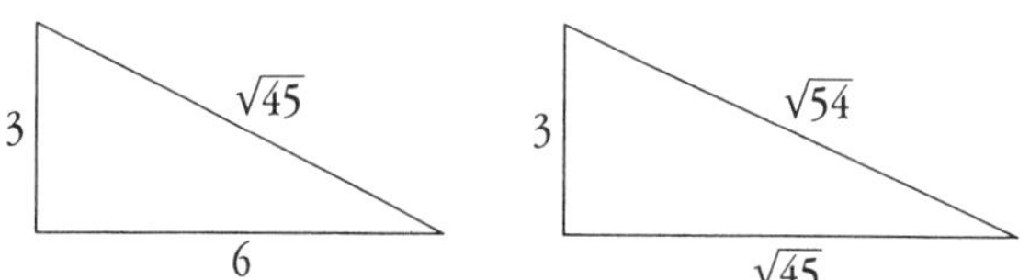

14.

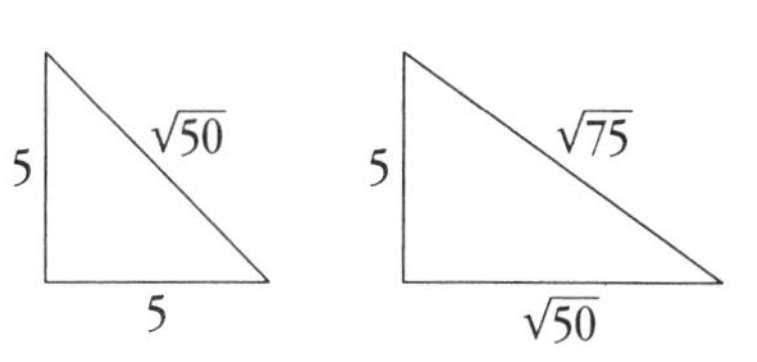

15.

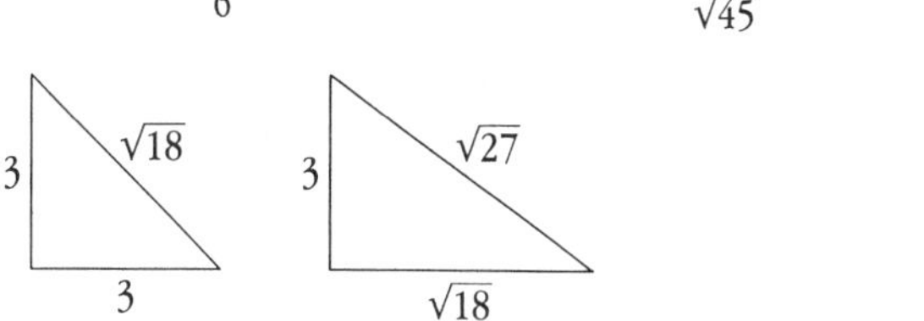

16.

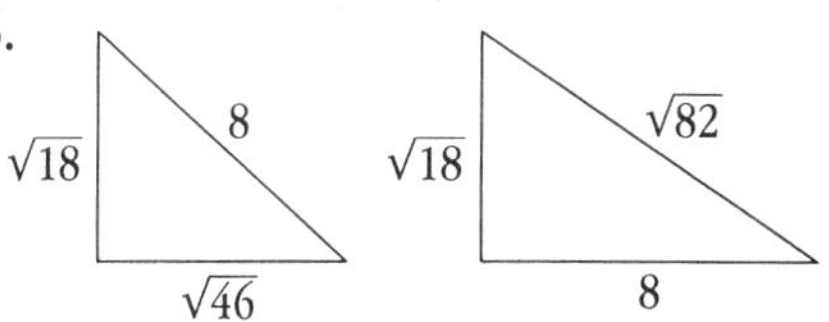

17.

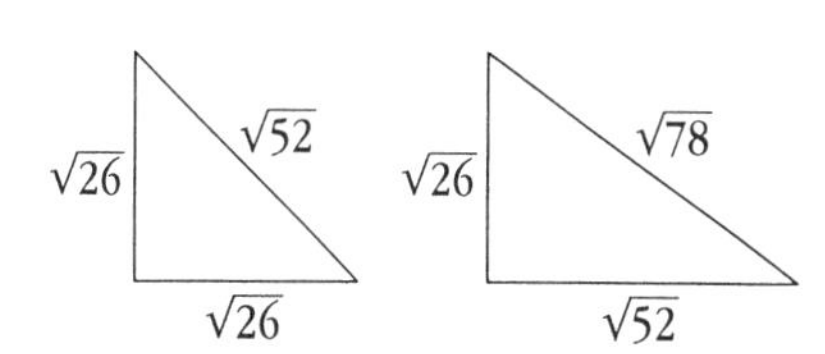

Investigation 5

1. a. 0.009009 . . . , 0.018018 . . . , 0.027027 . . . , 0.036036 . . .

   b. Possible answers: $\frac{1}{111} = \frac{9}{999} = \frac{6}{666} = \frac{3}{333}$, $\frac{2}{111} = \frac{18}{999} = \frac{12}{666} = \frac{6}{333}$, $\frac{3}{111} = \frac{27}{999} = \frac{18}{666} = \frac{9}{333}$, $\frac{4}{111} = \frac{36}{999} = \frac{24}{666} = \frac{12}{333}$

2. a. 0.030303 . . . , 0.060606 . . . , 0.090909 . . . , 0.121212 . . .

   b. Possible answers: $\frac{1}{33} = \frac{3}{99} = \frac{2}{66}$, $\frac{2}{33} = \frac{6}{99} = \frac{4}{66}$, $\frac{3}{33} = \frac{9}{99} = \frac{6}{66}$, $\frac{4}{33} = \frac{12}{99} = \frac{8}{66}$

3. a. 0.003003 . . . , 0.006006 . . . , 0.009009 . . . , 0.012012 . . .

   b. Possible answers: $\frac{1}{333} = \frac{3}{999} = \frac{2}{666}$, $\frac{2}{333} = \frac{6}{999} = \frac{4}{666}$, $\frac{3}{333} = \frac{9}{999} = \frac{6}{666}$, $\frac{4}{333} = \frac{12}{999} = \frac{8}{666}$

4. 4.777 . . .
5. 0.636363 . . .
6. 3.003003 . . .
7. 2.636363 . . .
8. 0.870870 . . .
9. 0.870870 . . .
10. $\frac{24}{99}$
11. $\frac{8}{99}$
12. $4\frac{32}{99}$
13. Possible answer: $\frac{36}{99} = \frac{12}{33} = \frac{4}{11}$
14. a. 4 and 5
    b. 9 and 10
    c. 15 and 16
    d. $\sqrt{400} = 20$
    e. 40 and 41
    f. Use the $\sqrt{\ }$ key to find a decimal approximation. The whole-number part will be the smaller of the two whole numbers. Add 1 to get the next largest whole number.
15. true
16. false
17. false
18. false

Investigation 6

1. The slopes of the segments are as follows: AL, $\frac{2}{5}$; BL, $\frac{2}{2} = 1$; KL, $-\frac{3}{1}$; LD, $\frac{2}{1} = 2$; IJ, $\frac{1}{2}$; JD, $\frac{1}{4}$; AJ, $\frac{3}{2}$; JE, $\frac{3}{4}$; GD, $-\frac{2}{6} = -\frac{1}{3}$; BD, $\frac{4}{3}$.
2. a. trapezoid
   b. isosceles triangle
   c. isosceles triangle
   d. right triangle
   e. scalene triangle
3. a. $10 + \sqrt{20} \approx 14.47$
   b. $2 + 2\sqrt{5} \approx 6.47$
   c. $4 + 2\sqrt{5} \approx 8.47$
   d. $5 + \sqrt{13} \approx 8.61$
   e. $5 + \sqrt{2} + \sqrt{17} \approx 10.54$
   f. $2\sqrt{8} + 2\sqrt{10} \approx 11.98$

Descriptive Glossary

hypotenuse The side of a right triangle that is opposite the right angle. The hypotenuse is the longest side of a right triangle. In the triangle below, the side labeled c is the hypotenuse.

irrational number A number that cannot be written as a fraction with a numerator and a denominator that are integers. The decimal representation of an irrational number never ends and never shows a repeating pattern of digits. The numbers $\sqrt{2}$, $\sqrt{3}$, $\sqrt{5}$, and π are examples of irrational numbers.

perpendicular Meeting at right angles. For example, the sides of a right triangle that form the right angle are perpendicular.

Pythagorean Theorem A statement about the relationship between the lengths of the sides of a right triangle. The theorem states that if a and b are the lengths of the legs of a right triangle and c is the length of the hypotenuse, then $a^2 + b^2 = c^2$.

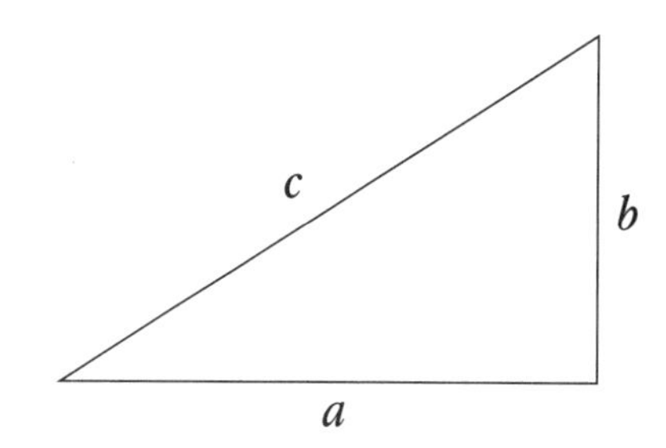

rational number A number that can be written as a fraction with a numerator and a denominator that are integers. The decimal representation of a rational number either ends or repeats. Examples of rational numbers are $\frac{1}{2}$, $\frac{79}{81}$, 7, 0.2, and 0.191919. . . .

real numbers The set of all rational numbers and all irrational numbers. The number line represents the set of real numbers.

repeating decimal A decimal with a pattern of digits that repeats forever, such as 0.3333333. . . and 0.73737373. . . . Repeating decimals are rational numbers.

square root If $A = s^2$, then s is the square root of A. For example, $^-3$ and 3 are square roots of 9 because $3 \times 3 = 9$ and $^-3 \times {}^-3 = 9$. The $\sqrt{\ }$ symbol is used to denote the positive square root. So, we write $\sqrt{9} = 3$. The positive square root of a number is the side length of a square that has that number as its area. So, you can draw a segment of length $\sqrt{5}$ by drawing a square with an area of 5; the side length of the square will have a length of $\sqrt{5}$.

terminating decimal A decimal that ends, or terminates, such as 0.5 or 0.125. Terminating decimals are rational numbers.

Index

Index

Index